Conceptual Design for Engineers

Second Edition

M. J. French MA MSc FlMechE
Professor of Engineering Design
University of Lancaster

The Design Council
London

Springer-Verlag
Berlin Heidelberg New York Tokyo

Conceptual Design For Engineers
Second Edition

First edition published 1971
as *Engineering Design: The Conceptual Stage*
by Heinemann Educational Books Ltd, London

Typeset by Sunrise Setting, Torquay

Printed and bound in the United Kingdom by
The Pitman Press, Bath

British Library CIP Data

French, Michael
    Conceptual design for engineers
    1. Engineering design
    I. Title
    620'.00425   TA174

ISBN 0 85072 155 5 The Design Council London
ISBN 3 540 15175 3 Springer-Verlag Berlin Heidelberg New York Tokyo
ISBN 0 387 15175 3 Springer-Verlag New York Tokyo Heidelberg Berlin

Contents

Preface to Second Edition   vii

Preface to First Edition   viii

Units and symbols   ix

Questions   ix

1   Introduction
1.1   Design: conceptual design: schemes   1
1.2   The anatomy of design   1
1.3   The scope and nature of design methods   4
1.4   Insight   5
1.5   Diversification of approach   6
1.6   Reduction of step size   7
1.7   Prompting of inventive steps   7
1.8   Generation of design philosophies   8
1.9   Increasing the level of abstraction   8
1.10 Design and the computer   10

2   Combinative ideas
2.1   Introduction   13
2.2   Construction of tables of options: functional analysis   14
2.3   Functional analysis: axial flow compressor rotor   15
2.4   Parametric mapping of viable options   17
2.5   Liquid natural gas tanker: alternative configurations   23
2.6   Further examples of combinative treatments   25
2.7   Elimination procedures for tables of options   28
2.8   Other ways of reduction   30
2.9   Synthesis from tables of options: kernel tables   32
2.10 Evolutionary techniques: hybrids   33
2.11 An example: wave energy converters   34
2.12 Evolutionary techniques: redistribution of functions   40
2.13 Repêchage   43
2.14 Combinative ideas: general remarks   44
      Questions   45
      Notes   45

3 Optimisation

3.1	Introduction	49
3.2	Making capital and running costs commensurate	49
3.3	Optimum speed of a tanker	50
3.4	The optimisation of the sag:span ratio of a suspension bridge	52
3.5	Optimisation with more than one degree of freedom: heat exchanger	55
3.6	Putting a price on heat-exchanger performance	57
3.7	Variation of costs with application	59
3.8	Further aspects of heat-exchanger optimisation	59
3.9	An elementary programming problem	60
3.10	Classification of optimisation problems and methods of solution	62
3.11	The design of rotating discs: an optimum structure	66
3.12	Hub design	73
3.13	Summary	73
	Questions	73
	Answers	74

4 Insight

4.1	Introduction	76
4.2	Rough calculations	76
4.3	Optimisation of compressor shaft diameter	83
4.4	The optimum virtual shaft: a digression	85
4.5	Useful measures and concepts	87
4.6	Bounds and limits	91
4.7	Scale effects	94
4.8	Dimensional analysis and scaling	98
4.9	Proportion	99
4.10	Change of viewpoint	100
	Questions	102
	Answers	104

5 Matching

5.1	Matching: the windlass	107
5.2	An extended example of matching: ship propulsion	107
5.3	Matching within a single machine	111
5.4	Further aspects of ship propulsion	112
5.5	Specific speeds: degrees of freedom	113
5.6	Matching of a spring to its task	115
5.7	Matching in thermodynamic processes	117
5.8	Two old cases of matching	121
5.9	Car handbrake	123
5.10	General remarks on matching problems	126
	Questions	126
	Answers	129

6   Disposition

6.1	Problems of disposition	132
6.2	Dovetail fixing	132
6.3	Splined shaft	133
6.4	Disc brake caliper: breaking logical chains	134
6.5	Alternator rotor	136
6.6	Structures: the feather	137
6.7	Structures: form design	139
6.8	Joints	141
6.9	The design of gear teeth	142
6.10	Disposition of allowable stress	146
6.11	Electromagnets	146
	Questions	147
	Answers	149

7   Kinematic and elastic design

7.1	Introduction	153
7.2	Spatial degrees of freedom: examples of kinematic design	153
7.3	Elastic design and elastic pairs	160
7.4	Comparison of kinematic and elastic means	166
7.5	Elastic design and structural economy	169
	Questions	171
	Answers	174

8   Costs

8.1	Design and costs: tasks, values and assets	178
8.2	Tasks and values in gearing	180
8.3	Gears for contra-rotating propellers	183
8.4	Large pressure vessels for gas-cooled reactors	184
8.5	Other cases	185

9   Various principles and approaches

9.1	Introduction	187
9.2	Avoiding arbitrary decisions: degrees of freedom of choice	187
9.3	Mathematical models	189
9.4	The search for alternatives	192
9.5	Logical chains	193
9.6	Past practice and changed circumstances	195
9.7	Brainstorming	196
9.8	Use of solid models	196
9.9	Some maxims for designers	197
	Questions	200
	Answers	201

10 Conclusion

10.1 Resumé   203
10.2 Checking and evaluation   203
     Questions   205
     Answers   210

     References   219
     Index   222

Preface to the Second Edition

At the time this book first appeared the importance of design in engineering education was not widely accepted: above all, it was often seen as a thing apart, to be taught without close links with the analytical parts of the course. Attitudes have since changed, and it is now recognised that design should have a central and integrating role: as it was put in the Design Council report on *Engineering Design Education* (The Moulton Report), 'Engineering should be taught in the context of design, so that design is a continuous thread running through the teaching of undergraduate engineering.' This view has since been widely endorsed, most recently in the Report of the Working Party of the Science and Engineering Research Council on Engineering Design (The Lickley Report).

It is this climate of opinion above all that makes it seem timely to re-issue this book, which should be an ideal text for second and third year mechanical engineering students. Showing as it does how relatively simple engineering science, along with commonsense considerations, dominates the early or conceptual stages of design, it can help them achieve the kind of thinking, synthetic rather than analytical, that distinguishes the engineer. Staff, too, will find the book a source of ideas for mind-stretching exercises which are, however, still within the scope of students with little practical experience. The numerical or algebraic nature of many of the examples helps to dispel the idea that design is lacking in rigorous lines of thought.

The opportunity has been taken in this new edition to make various changes, chiefly the substitution for old examples of ones which are more recent, of wider interest or simply easier to follow. Some material of doubtful value has been dropped, and a few new ideas have been added, particularly some maxims, which encapsulate much of what is said elsewhere.

M. J. French
Lancaster July 1984

Preface to the First Edition

This book is about the early or conceptual stages of design—more specifically engineering design, but many of the ideas put forward are equally applicable to all kinds of functional design and so will be of interest to industrial designers and others.

Much of the book requires little or no engineering knowledge, and all of it is within the grasp of a second-year engineering student. The demands made on the reader's knowledge are small, but those made upon his intelligence are less so.

It is hoped that reading this book and wrestling with the problems in it will be of great benefit to young engineers, not only those who later become designers but also those who will be involved in creative engineering in other roles, such as research and development, or who will supply specialist functions like stress analysis in support of design. It should also help those who, as managers, will have to make decisions about design policy.

Further, it is felt that the practising designer, to whom most of the ideas will be at least somewhat familiar, will find this book of interest and perhaps of value in strengthening and refining his approach. And if one such, reading it, says here, 'I know a better example than this,' and there, 'But hasn't he realised you can look at such problems so?,' and sits down and writes a better book, that will be so much the more justification for this one.

It is not the purpose of this book to add to the reader's stock of theory or useful facts; its purpose is to cultivate an attitude of mind, a discipline of thought, and a set of approaches to design problems. To this end, many peripheral topics, such as the management of design and the relationship of design to other functions, have been left out. Many books, ostensibly on the same subject, deal almost entirely with these fringe activities, consigning the central issue to a blackbox mysteriously inscribed 'devise solutions,' or something of the kind: this fault at least has been avoided.

While not quite a textbook, this book should make good background reading for engineering students, and engineering teachers will find it contains much matter on which exercises for design courses can be based. There are over 70 problems with sketch solutions and notes, on a very wide range of topics.

To economise in background explanations, the same example has often been used several times to illustrate different points. No apology should be needed for this device, since it reflects the value in design, as in other fields, of looking at the same thing from several points of view.

M.J.F.
Lancaster 1971

Units and symbols

Units

Mostly SI units are used. In a few cases British units or both kinds are given. For reference, unit abbreviations which may be unfamiliar are listed here.

A	ampere	N	newton $(= \mathrm{kg\,m/s^2})$
ft	foot	s	second
in	inch	W	watt $(= \mathrm{joule/s})$
J	joule $(= \mathrm{newton\text{-}metre})$	Wb	weber
K	degrees Kelvin	p.s.i.	pounds force per square inch
kg	kilogram	p.s.i.a., p.s.i.g.	pounds per square
kWh	kilowatt hour		inch absolute and gauge
lbf	pound force	SHP	shaft horsepower
m	metre		

For the convenience of those not yet familiar with the newton, $1\mathrm{N} \simeq 0.2248\,\mathrm{lbf}$. Pressures and stresses are given mostly in newtons per square millimetre and $1\mathrm{N\,mm^{-2}} \simeq 145.0\,\mathrm{p.s.i.}$

Symbols

Because of the large variety of topics and the brevity of their treatment, notation has been introduced and explained in the text. Customary usage has been followed in appropriate cases.

Questions

A number of questions (Q) are given at the end of each chapter, except Chapters 1 and 8, together with outline answers (A) or notes (N) on their solution. The average standard is rather high, and it is not expected that the reader will be able to solve many of them without the help provided. In some cases the question number is followed by a number between 1 and 5 intended to indicate difficulty; e.g. Q.4.5(3) indicates the fifth problem on Chapter 4 and that it is of the third order of difficulty.

The questions are mostly intended as worked examples forming an extension of the text, throwing new light on the subject matter and indicating ways of applying the methods and ideas advanced, so that the reader who wishes to go more deeply into the subject should certainly study them. The questions of Chapter 10 cover matter from throughout the book.

1 Introduction

1.1 Design: conceptual design: schemes

In this book 'design' is taken to mean all the process of conception, invention, visualisation, calculation, marshalling, refinement, and specifying of details which determines the form of an engineering product.

Design generally begins with a need. This need may be met already by existing designs; in such cases the designer hopes he can meet the need better (i.e. generally more cheaply). It ends with a set of drawings and other information to enable the thing designed to be made. Ideally the intervening stages should be of successively increasing precision, of gradual crystallisation or hardening. The early stages differ in character somewhat from the later ones, largely because of the greater fluidity of the situation. These early stages, when there are still major decisions to be made, are called 'conceptual design' [1]. The products of the conceptual design stages will be called 'schemes'.

By a scheme is meant an outline solution to a design problem, carried to a point where the means of performing each major function has been fixed, as have the spatial and structural relationships of the principal components. A scheme should be sufficiently worked out in detail for it to be possible to supply approximate costs, weights, and overall dimensions, and the feasibility should have been assured as far as circumstances allow. A scheme should be relatively explicit about special features or components but need not go into much detail over established practice.

1.2 The anatomy of design

Figure 1.1 is a block diagram showing the design process according to the writer. The circles represent stages reached, and the rectangles represent work in progress.

Constructing block diagrams is a fashionable pastime, especially in fields like design where boundaries are imprecise and interactions legion, so that any ten experts will produce ten (or a hundred). They will all be different, and all valid— 'there are nine and sixty ways of constructing tribal lays, and every single one of them is right.' They express only truisms, and yet they have a value for all that.

In this particular block diagram things that are important but outside the scope of this book have been omitted—for example, relationships with other activities like

1

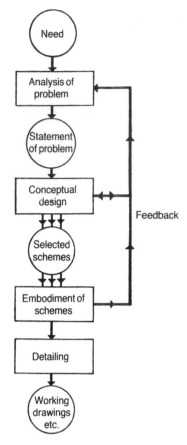

Figure 1.1   Block diagram of design process.

research and development, inputs of information and so forth. On the other hand, there is no box labelled 'evaluation' because the writer believes it should be going on continuously in all the rectangles.

Analysis of the problem

This part of the work consists of identifying the need to be satisfied as precisely as is possible or desirable. For such things as consumer goods or widely used items of machinery, it is a problem of locating and defining an adequate 'ecological niche' and this is the hardest case.

   The analysis of the problem is a small but important part of the overall process. The output is a statement of the problem, and this can have three elements:

   (1) a statement of the design problem proper,
   (2) limitations placed upon the solution, e.g. codes of practice, statutory requirements, customers' standards, date of completion, etc.,
   (3) the criterion of excellence to be worked to.

Now the ultimate criterion of excellence is always cheapness; all such things as reliability, efficiency and so forth can be reduced to costs, given sufficient information, and always should be as far as possible. Even matters of safety ought to be examined on a basis of cost-effectiveness. The best design is the cheapest, provided that the cost is properly assessed.

However, this ultimate criterion of low cost is often inconveniently remote, and more immediate criteria are sometimes of greater use. For example, in designing an aircraft gas turbine it is useful to know the equivalent capital value of saving a unit of weight so that a limit is set within which the lightest design is the cheapest.

An important point to notice about the analysis of the problem is that there is feedback to it from the design work proper. Few needs are absolute; most are relative to the costs of filling them. The design of a private car, for example, might well develop in such a way as to change its original statement of problem. Features which were excluded as likely to make the price too high might prove cheaper than was expected, and others originally regarded as essential might be so irreconcilable with the developing design philosophy (see Section 2.9) as to become unduly expensive.

Conceptual design

This phase of the design process forms the main topic of this book. It takes the statement of the problem and generates broad solutions to it in the form of schemes. It is the phase that makes the greatest demands on the designer, and where there is the most scope for striking improvements. It is the phase where engineering science, practical knowledge, production methods, and commercial aspects need to be brought together, and where the most important decisions are taken.

Embodiment of schemes

In this phase the schemes are worked up in greater detail, and if there is more than one, a final choice between them is made. The end product is usually a set of general arrangement drawings. There is (or should be) a great deal of feedback from this phase to the conceptual design phase, which is why the writer advocates overlapping the two.

Detailing

This is the last phase, in which a very large number of small but essential points remain to be decided. The quality of this work must be good, otherwise delay and expense or even failure will be incurred: computers are already reducing the drudgery of this skilled and patient work and reducing the chance of errors, and will do so increasingly.

This brief description of the design process will suffice for present purposes. Apart from organisational problems, the difficulty of maintaining the quality of detail draughting, and so forth, there are three central problems in design:

  (1) the generation of good schemes (conceptual design),
  (2) securing the best embodiment of those schemes (the problem of best embodiment), and
  (3) the evaluation of alternatives.

This book is primarily about the first problem, but inevitably becomes concerned with the others. Also, many of the approaches suggested for generating schemes are directly applicable to the embodiment stage as well, since the design of a part may be tackled much as if it were a complete design problem.

1.3   The scope and nature of design methods

This book is an attempt to show what design methods can do to help the designer, principally in the conceptual stage: it will no doubt appear very amateurish in future years when the subject has been thoroughly developed.

First, let us be clear what design methods cannot do and probably never will be able to do. They cannot replace the gifts of the talented designer, nor provide step by step instructions for the production of brilliant designs. What they may perhaps do is to improve the quality and speed of the able designer's work, and increase the size and range of the tasks he can tackle. They may also speed the development of the young designer and, perhaps most important of all, improve the co-operation with essential specialists inside and outside the design office.

The ways in which design methods may help can be roughly classified as follows:

(1) by increasing insight into problems, and the speed of acquiring insight,
(2) by diversifying the approach to problems,
(3) by reducing the size of mental step required in the design process,
(4) by prompting inventive steps, and reducing the chances of overlooking them, and
(5) by generating design philosophies (synthesising principles, design rationales) for the particular problem in question.

Other benefits may be expected from the keeping of better records of the progress of design work, from wider-spread and more accurate knowledge of the reasons for past decisions, even from less ambiguous and more rapid communication between designers.

Not all design methods will be useful when applied to any given problem: an analogy may help to explain this. Consider a boy trying to prove a rider in Euclidean geometry. He knows a number of theorems, and he has to find those which will help him in his particular problem. The problem itself provides some clues—there is a circle with a tangent and a chord, so he applies the theorem about the angle between a tangent and a chord. This probably does not solve his problem, it merely presents him with a changed situation that may suggest a further step. The kind of talent the boy requires in applying theorems is rather like that which the designer needs in applying methods. As with the designer, there are generally many solutions to the boy's problem (which is to find a *proof*, not a result), some elegant, others clumsy but adequate. Like the designer, he will feel at times he is making progress, and sometimes the feeling will be justified and sometimes it will not.

Design problems, like other tough creative problems, are solved by sheer hard mental work. Methods help the designer to keep himself working, to tide over usefully those periods when inspiration will not come, or to break out of circular arguments. They help him to grasp the eel-like tail of a solution as it flashes through the corner of the imagination and pin it out on the drawing board, when, alas, it often turns out to be unworkable. Even then, however, method will often mean it is more quickly seen to be useless.

The methods that are discussed in the rest of this book, which are sometimes called ideas, approaches, techniques, or aids, according to their degree of formalisation or the width of their applicability, have certain things in common. They involve a removal of the problem to a higher level of abstraction; an attempt to clear away the purely circumstantial and to look for a general element which will widen the field of known means in which a solution can be sought; an attempt to substitute measures for merely qualitative words like advantage and disadvantage, dear and cheap; in short, the sort of bricks with which reason has always built.

1.4   Insight

It is a common experience to find that an idea, worked on long and hard, has some basic flaw which is instantly recognised at a later stage because insight into the problem has improved. It is therefore wise for the designer tackling a new problem to spend some of his time improving his insight into it as rapidly as possible. This will not be done by careful studies along one set of lines, but by many rapid exercises, looking at different aspects in different ways, a sort of familiarisation procedure.

The ancient problem illustrated in Figure 1.2(a) may serve to illustrate the value of insight and how it may be gained by looking at the problem in a different way. The black knights and the white knights have to be exchanged, subject to the ordinary rules of chess for moving knights, by means of the minimum number of journeys, where a journey is any number of consecutive moves by the same knight. Now if we remove three knights and move the remaining one continually, never going back to the square just left, it traces out a closed path. If we start with the black knight on the square numbered 1 and move him to 2, his only move then, apart from going back, is to 3, and then to 4, 5 and so on. If we unfold the twisted loop of Figure 1.2(a) we have the entirely equivalent arrangement of Figure 1.2(b) where a knight's move is one space forwards or backwards round the circle. It is now easy to see that the minimum number of journeys is seven (say, black knight on 1 to 2, white knight to 1, white knight to 8, black knight to 7, black knight to 5, white

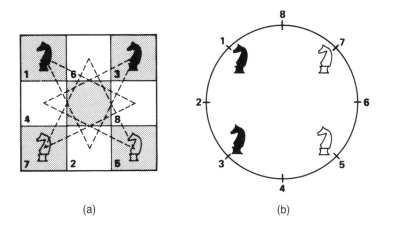

(a)   (b)

Figure 1.2   Knights problem.

knight to 3, white knight to 1, all anticlockwise) and that the task can be performed in just eight different ways, one starting with each of the eight possible first moves.

Notice how the clue to the simpler viewpoint was afforded by studying the trivial problem of how one knight can move on the empty board.

Here is an engineering example of the importance of insight. It was suggested that a special tanker for carrying liquid natural gas at about $-160\,°C$ should have no insulation provided, but would generate its own in the form of a thick layer of ice which would rapidly deposit on the hull. Quite apart from the objections to a cold hull, this suggestion can be dismissed out of hand: it does not matter how thick the ice is, its surface cannot be above about $0\,°C$ and the surrounding sea will average, say, $15\,°C$. With this temperature difference and the water racing over the surface of the ice at perhaps 17 knots, the rate of heat-transfer would be much too high and the cargo would rapidly boil away. It would be a waste of time even to look up the equilibrium temperature of ice in sea water.

1.5   Diversification of approach

A great resource when faced by a difficult problem which shows no sign of yielding is to attack it from another side or in another way. This is, of course, as well understood by designers as anyone else, but it is not always so easy to find another way, and this is where the sort of ideas discussed in this book may help.

Suppose someone asks us how to go from A to B, a journey we have never made in an area where we are well acquainted with all the roads. We can answer much more readily with a map before us, though it does not add anything essential to our knowledge. We can at once see all the alternative routes which otherwise would take time to think out, and all the considerations like level crossings or congested areas which influence the choice between them and might otherwise be forgotten. So design methods can prompt ways of tackling problems which otherwise might not be hit upon or, more probably, hit upon so late that the work they would have saved has already been wasted.

Very often, the designer has difficulty in bringing his mind to bear profitably on the problem in hand: release from this unhappy state is often brought by a happy thought, a minor inspiration, which causes him to readjust slightly his point of view, and progress is resumed. One of the ideas in this book might well have brought that happy thought at less cost in time and brain-cudgelling. It has often been suggested that such seemingly unprofitable hours are a necessary precursor of the leap forward (e.g. reference 2) but the writer does not believe this is true of the small steps that have to be made every day.

Viewed from a single aspect, a design problem often shows too great a variety of possibilities: walk around it, and certain imperatives are borne in on us, some constraints which restrict the areas in which a solution may lie. Design is a catch-as-catch-can business—we must seize a problem wherever there is something to take hold of. For example, in a particular heat-exchanger problem, where the range of forms which might have proved cheapest appeared at first sight very large, the consideration of what the baffles had to be like and the resulting difficulties of assembly pruned the possibilities much more satisfactorily than any other aspect.

1.6   Reduction of step size

Some great ideas appear so obvious it is difficult to understand why they were not thought of earlier. Sometimes a contributory reason for this is that hindsight has, as it were, a series of unacknowledged stepping stones whereby to cross the stream that inspiration has leapt. The great idea can be explained in a series of steps, none of which in itself appears remarkable, and once such an explanation has been given, it is difficult ever again to grasp how wide the gap was. The moral is that those of us who are not good long-jumpers should look for handy stones when we wish to cross streams.

Among the small steps into which the arguments leading to good and profitable design innovations may be broken, we may expect some to be rather special to the problem in question, but others to be of a more general kind, the sort that might be made as part of a routine approach. Various steps of this sort occur in the examples discussed in this book.

One elementary but useful way of reducing step size may help to give the general idea. If a difficult function has to be performed, we can make some progress by listing the means we can think of which meet some of the requirements and then tackling the separate smaller problems of adapting each of these to meet the full requirements: with any luck we shall be able to solve one or more of these smaller problems, perhaps by introducing an *auxiliary function*, or by recognising that the original function may be divided into two (see Section 2.12). To give a trivial example, we might need to support a large load in a region at a high temperature, and this might be done by using a member suitable only for a lower temperature and surrounding it by a refractory shroud with cooling air flowing through it.

1.7   Prompting of inventive steps

The claim that method may prompt inventive steps will seem rash, if not ridiculous, to some. But arguments which can be built on the lines put forward here will often reduce to a marching logic which leads inexorably to a minor but unmistakeable invention: all this, it seems to the writer, can happen without any very great mental leaps.

It is difficult to illustrate this. The field in which the writer has been most aware of this possibility is that of thermodynamic process design, where the problems are relatively tidy and precise, the available means strictly limited and the rules of the game well understood and immutable. In this specially favourable case it is reasonable to imagine a computer, as it were, nudging the designer and saying 'Look at this, look at these particular values, shouldn't there be an improvement possible here?' Indeed, the writer has had patented ideas produced in just this way, except that he had to go through the figures himself to find the prompting values.

The Linde–Fränkl process for the separation of atmospheric oxygen and nitrogen is an example of marching logic leading to a major inventive step in the thermodynamic process field[3].

1.8   Generation of design philosophies

It is a characteristic of much good design that it has a consistency and internal coherence, a quality that the Germans call '*Eindeutigkeit*' but might be called 'clarity of function'. Often this springs from a central informing idea, or design philosophy. Examples will be given later, particularly in Chapters 2 and 9, but the famous Moulton small-wheeled bicycle (1962) will illustrate the point[4].

Moulton began by querying the need for large wheels on a bicycle. Large wheels were less compact, heavier, less stiff and produced more air drag than small wheels, which would seem to have all the advantages. But if small wheels had the same section of tyre as large ones they would have to be inflated to a much higher pressure because of the short contact patch on the road, and they would then give a rough ride. If, on the other hand, they had a large-section tyre inflated to a lower pressure, the ride would be comfortable but the rolling resistance would be high, and the cyclist would have to work harder than necesssary. Moulton decided that his bicycle should be small-wheeled, of low resistance and comfortable, and the only way that could be done was by using narrow, high-pressure tyres and a separate suspension to give comfort. From this starting point the whole design developed.

Recently (1983) Moulton has launched a later version, differing chiefly in the form of the front suspension and the structure of the frame, but retaining the essential design philosophy, which may be summed up as providing the advantages of the small wheel without its possible disadvantages. What is not apparent from such a brief description is the elegant 'clarity of function' of the design that this philosophy informs (Figure 1.3). In this latest version the bicycle separates in two for convenient stowing by a joint in the middle of the space frame of slender tubes which unites seat pillar and steering column, a beautiful solution reached after considering and rejecting many slightly unsatisfactory alternatives.

Like the Moulton bicycle, many designs have their basis in a few key decisions, arising from a few important considerations, out of which a great many consequences flow. It is convenient to call such a basis a 'design philosophy' and once it has been arrived at, it often produces a steady flow of secondary decisions which rapidly map out the whole course of the work, sometimes via a step-by-step process which has been called 'marching logic'. However, it is not easy to light on such an informing principle, and frequently it emerges only when the scheme is largely complete, and its value is then small. One of the objects of the approaches discussed later is to try to hasten the appearance of such a design philosophy.

1.9   Increasing the level of abstraction

Two simple precepts the designer should always bear in mind are 'quantify whenever possible, even if only roughly' and 'increase the level of abstraction at which problems are treated until it ceases to be profitable'. Both these precepts are extensively illustrated later. The second, however, warrants some enlargement at this point because it is an important characteristic of almost all the ideas put forward.

Figure 1.3 Moulton bicycle.

All mental disciplines must rest heavily on the use of abstract concepts, which are essential to intellectual processes. Moreover, progress in understanding is often associated with increasing levels of abstraction. But abstraction may take different forms, and it is important in design that the right forms should be selected, especially as specialist writings in the engineering field are mostly analytical. As a result there is a grave shortage of good creative abstractions; those that exist are not widely known, and accepted terminologies are virtually non-existent.

As an example, it seems unlikely that many readers are familiar with the term 'corner power' (see Section 4.6), and yet this is a most helpful concept in many fields of design. Sensible heat (enthalpy rise associated with temperature rise) has been banished from the thermodynamic vocabulary, perhaps rightly, but no substitute has appeared. The analyst may be able to avoid the use of such a term without much circumlocution, but for the designer the loss is serious. Many thermodynamic processes suffer from what the writer calls 'sensible heat cycling', where a large 'water equivalent' (another abstraction, which is useful in design and is falling into disfavour) is subjected to large temperature cycles. Sensible heat cycling nearly always involves large losses as, for example, in Newcomen's engine before Watt improved it (Section 2.12). Stirling-cycle engines are a rare example where sensible heat cycling is nearly reversible. It is a great impediment to the discussion of the design of a thermodynamic process not to be able to say something like: 'but if we do that we shall increase the sensible heat cycling', and be understood.

Another useful concept is 'holding-together power', the product of the tensile stress and the volume in a tensile structural material. This, like corner power, is an example of a measure of a task to be performed or the ability to perform a task. If a pressure vessel has a volume V and has to stand an internal gauge pressure p then its holding-together power is $3pV$ (see Section 8.4).

More abstract does not always mean more general. If we want to design an elastic beam, the highly abstract but very specialised view of the beam as two flanges and a web, the flanges taking all the moment and the web all the shear, is immeasurably more useful than the very general theory of elasticity. The key to the superiority of the cruder concept here is its greater abstraction (it has only three areas and a depth, at most, defined at any section) and its *purpose-related* nature. The beam has a *resistant function* (Torroja), which is to resist a certain pattern of shear and bending moment over a certain length. We provide separately for the two parts of its function, a web, to resist shear, and two suitably spaced flanges to sustain the bending moment.

The kinds of abstraction specially useful in design are generally those associated with purposes, with tasks or the ability to perform tasks, with limitations or with losses.

1.10   Design and the computer

The computer has already had a large influence on engineering design, and this influence can be expected to extend. So far its chief contribution has been at the detailing stage, where its use is becoming general. In companies producing large numbers of similar designs, often differing only in a few parameters in the specification, much of the work has been done by computers for a generation. A good example is provided by electrical power transformers, where the specification contains little more than the voltage, the power, the frequency (50 or 60 Hz) and some rating and environmental particulars. Design consists largely of deciding which frame size, standard lamination and so forth will meet the specification at least cost and determining the leading dimensions, and this was being done by computer in the days before transistors, by machines using thermionic valves. Nowadays the computer carries out, not only these tedious and repetitive calculations, but also the preparation of manufacturing instructions for the machines making many of the parts. An important commercial aspect of these developments is the speed and confidence with which tenders can be made and the short delivery times which can be offered. There are a number of such major components, made by specialist manufacturers on a design and supply basis, which are so treated—electric motors, heat exchangers, pumps and so on.

Computers have been used in this way for a long time in shipbuilding and the design of process plants where there is much routine detailing and the programs for computer-controlled manufacture can be generated at the same time.

Computer-aided draughting and computer-aided manufacture are becoming the rule everywhere. But the purist will object, quite rightly, that these are not design, and certainly not conceptual design. Computers have not yet reached the stage at which they can contribute to the essential design process (though one possible use verging on design is mentioned in Section 1.7, and there are others) but some recent developments have brought nearer the day when they will. Meanwhile the modern designer depends on the computer to perform the calculations which guide his work. For example, finite element methods are nowadays the routine way of stress analysis for all but the simplest or least important mechanical engineering structures. The design of a car body may proceed by the designer drawing something he guesses is about right, based on experience and insight, which is then analysed. If it is too flexible or overstressed, the designer will then make changes, the analysis will be repeated, and so on (Figure 1.4). There are programs in which scantlings are automatically adjusted to meet requirements and to minimise weight, but these are confined to minor changes which do not change the structural layout.

Besides stress analysis, there are many other areas of engineering science in which the computer provides close technical support to the designer, sometimes interactively so that he can determine very quickly the effect of changes.

Simulation

One important use of the computer in design is in the simulation of complex dynamic systems. For example, the wavepower device described in Chapter 9 has a complicated dynamic behaviour and is subject to a very irregular forcing from the waves. While our understanding of such systems is quite well developed, it is a great benefit to have the computer, with which we can model mathematically the behaviour of the device in irregular seas, using data obtained from recording buoys at sea, and run 'months', if we wish, of 'tests', and analyse the results. In a similar way, real vehicles may be driven over rough roads to measure the dynamic inputs at the wheels, and these data may be used to compute the behaviour of an entirely new vehicle, one still on the drawing board. Perhaps the most amazing of all simulations is the study of supersonic flow past bodies using computer simulation to replace the

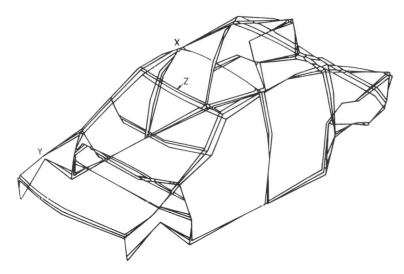

Figure 1.4   Finite element analysis: computed first bending mode, British Leyland Energy Conservation Vehicle. (By kind permission of the Institution of Mechanical Engineers, BL and Mr C. S. King.)

traditional experimental method with a model in a wind tunnel, a development made possible only by the great speed of modern computers. It is interesting to recall that simulation was once the field in which the now obsolescent analogue and hybrid computers were important, because, although their accuracy was inherently rather limited, they were relatively fast.

Besides this close technical support, there is another way in which the computer already aids the designer, by providing insight into form through three-dimensional graphics. Forms defined on the computer can be displayed from any angle, rotated, enlarged or contracted, at will. This facility is particularly important where the aesthetic aspect is dominant, but it is useful in purely functional design as well.

In addition, the computer (sometimes in a very rudimentary form) may actually form a component in a design. The microchip is cheap and varieties can be made to special requirements at a low cost in quite modest numbers, say 100 000. It is thus eminently practical to apply them in consumer goods such as washing machines, cameras and cars. Here they perform part of the task which was previously done by ingenious mechanisms, and they do it much better, at lower cost, in smaller space, and with greater reliability, so much so, that if there is a failure, it is usually in the electromechanical output devices. Many options open to designers in the past must have been rejected because the control functions required were of unmanageable complexity, requiring a forest of cams and gearing to achieve the necessary relationships: nowadays a microchip will replace most of the mechanism, while placing few demands on the designer's ingenuity.

Shortly, with more powerful computers and more advanced programs, the help that can be provided to the designer will be greatly extended. It will be possible, for example, to check whether components can be assembled, a task which sounds easy but is not easily converted into a 'computable' form. With this and much similar work done by the computer, the designer will have more time for the creative side.

2 Combinative ideas

2.1 Introduction

A design problem might be tackled on the following lines: take all the possible elements that might enter into a solution, combine them in all possible ways, and then select the best combination.

This ideal combinative process has three important characteristics: the synthesis aspect, such as it is, is reduced to a mechanical process of systematic combination, and the difficulties are reduced to those of analysis; it appears impossible to overlook good solutions; lastly, it is impracticable.

The chief reason for the impracticability of this comprehensive approach is the overwhelming number of possibilities inherent in almost all design situations. Put another way, the number of degrees of freedom of choice is much too large for all variations to be studied on their individual merits. There is some analogy with certain games. In noughts and crosses all the possible plays may be traced to their conclusion and the best play in any position determined. It is a game which can be treated exhaustively, or is 'exhaustible' for short. The writer is not expert enough to be sure if draughts is exhaustible—it is near the borderline perhaps—and chess, to date, is far from exhausted, notwithstanding the use of computers. Practically speaking, design problems are inexhaustible.

Nevertheless, much advantage may be drawn from the central idea of a combinative procedure, used sparingly in any of a number of different ways. As a very simple example, consider the design of a private car. We can put the engine at the front, F, or the rear, R, and we can drive the front wheels, f, or the rear wheels, r, giving four possible arrangements

Ff Fr Rf Rr

The principal advantages of these different arrangements are set out in Table 2.1. It can be seen that every scheme is superior to every other scheme in one or more respects, except Rf which has only one of the three advantages possessed by Ff and no other advantage. Rf may thus be regarded as wholly inferior to Ff and can be rejected, without attempting even to put numerical values to the various conflicting advantages. This is in accordance with historical practice, in which there are plentiful examples of the other three configurations but none of Rf. To proceed farther than this one easy discard, very detailed studies are needed. In the event, the Rr combination, suffering as it does from two disadvantages in road holding, has fallen into disfavour.

Before discussing the uses of combinative procedures in conceptual design we shall examine the resources they offer.

Advantage	Combination			
	Ff	Fr	Rf	Rr
Less flexure to be accommodated in shafts driving wheels		✓		✓
Better road holding due to forward C of G	✓	✓		
Better road holding due to front wheel drive	✓		✓	
No propellor shaft, clear floor space	✓			✓

Table 2.1   Table of options.

2.2   Construction of tables of options: functional analysis

The essence of all combinative methods is some sort of table of options, of which Table 2.1 is a very small example. Since the number of different combinations is often too large for each to be represented by a separate pigeonhole, the table is usually condensed by displaying only the separate options.

Various kinds of option generally exist, and they may be viewed in different ways, so that it is possible to construct tables on several bases, of which the chief are:

(1) alternative means of performing essential functions,
(2) alternative solutions to various smaller design problems or *sub-problems* [46] arising within the main one,
(3) a classification (virtually, a taxonomy) derived from a study of known solutions of the same or allied problems,
(4) alternative spatial arrangements, configurations or orders (Table 2.1 is an example),
(5) mixtures of some of the first four.

The choice of basis must depend on the nature of the problem and the preference of the designer.

Basis (1) is suitable for all new problems where no extensive background exists, but it can also help to throw new light on old problems.

Basis (2) is well adapted to cases with an extensive background, where the difficulties are well understood and the dominant sub-problems have emerged clearly. It is economical of effort in that it concentrates attention on the vital areas. Care must be taken, however, that the scope of enquiry is not limited to the point where radical alternatives are overlooked.

Basis (3) is of fundamental interest, in that the systematic tool is generated out of the results of non-systematic design. It can be used in very difficult cases in which it would be hard to make headway by any other procedure, and the table it produces is automatically well suited to the task in hand.

Basis (4) is most likely to be of use in parts of problems, or for studying a single aspect of a large problem.

2.3   Functional analysis: axial flow compressor rotor

An important tool for constructing tables of options is a *functional analysis*, which sets out the essential functions that the design is required to perform.

Consider the problem of the mechanical design of the rotor for an axial flow compressor, all the blade dimensions being already determined; such rotors are shown in Figures 2.2–2.6.

The principal functions to be filled by the rotor are:

(1) providing an attachment for the rotor blades,
(2) transmitting the drive to the rotor blades,
(3) resisting centrifugal loads due to the rotor blades,
(4) fairing the inner wall of the gas duct,
(5) providing adequate transverse stiffness.

The even-numbered functions here will be largely ignored. Although essential, (2) is a trivial problem to provide, and (4), though it has an important bearing, would complicate the example too much if properly taken into account.

The loads involved in (3) are very large relative to the size of the rotor, and make both (1) and (3) exacting problems. A large transverse stiffness (5) is needed to avoid shaft vibration or whirling troubles; we will adopt here the basically unsound but practically justified procedure of ensuring a minimum lowest transverse natural frequency of vibration of the rotor in rigid bearings as a precaution against whirling.

In forming a table of options for an axial flow compressor rotor design, functions (2) and (4) will be omitted. We proceed to find all the likely means we can of performing the remaining functions. It is unlikely that any new means will be hit upon—all the options that go into the table will probably come from what the Patent Office calls 'known means'.

For example, the attachment of the blades to the rotor, which for simplicity is assumed to be demountable, can be done by the well-known bulb, dovetail or fir-tree fixings, or by pins or rivets, or by circumferential grooves, as illustrated in Figure 2.1.

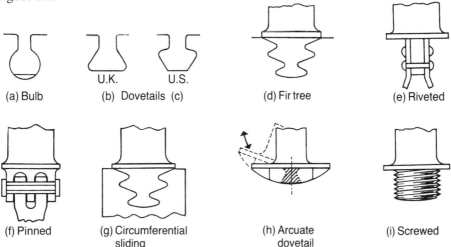

| (a) Bulb | (b) Dovetails (c) | (d) Fir tree | (e) Riveted |
| | U.K.   U.S. | | |

(f) Pinned   (g) Circumferential sliding   (h) Arcuate dovetail   (i) Screwed

Figure 2.1   Forms of blade root fixing for axial flow compressors.

Classification of blade attachments

It is worth making a small digression at this point to study the way in which a classification can be made of blade attachments; it brings out several points of interest.

One broad division is into those solutions that require extra parts and those which do not. The former require extra load transfers from part to part, and so, on a task-cost basis (Section 8.1), they must carry an inherent weight penalty, which may, however, be offset by higher working stresses or which may be justified by advantages such as reduced vibratory stress levels. Those which do not have extra parts have sliding engagement (another kind, having convergent engagement, is rare and unlikely, e.g. taper pipe threads and carpet sweeper handles), and so their mating surfaces must be composed of lines whose geometric properties are identical at all points in their length. Such lines must have constant curvature and torsion (rate of rotation of the plane of curvature about the tangent to the line), i.e. they must be helixes. A helix with zero torsion is a circle, and a helix with zero curvature is a straight line, giving the three kinds of sliding engagement associated with the three lower kinematic pairs; the screw pair, the turning pair, and the sliding pair. This is exhaustive—there are no other kinds.

It is useful to note in listing alternative blade attachments such characteristics as the effect they have on possible rotor constructions—straight-line sliding roots are usually impracticable with solid rotors—or the amount of dismantling necessary to change blades.

Function (3), resisting centrifugal loads, is treated exhaustively in Section 3.11. Here we shall consider three options, the disc, the ring or drum, and the solid rotor, the disc being about twice as economical in material as a ring or drum of thin radial extent.

Function (5) is similar to that of a bridge, and any form of bridge structure that is independent of anchoring and abutment thrust requirements would serve, but in practice only a beam or shaft is reasonable. One important choice that does remain is whether the shaft is a continuous piece, i.e. a separate shaft, or a stack of pieces butted end to end and prestressed by a central bolt or tendon or a ring of bolts, an arrangement which will be called a 'virtual' shaft. Figure 2.4 shows such a rotor construction. Finally, a drum or a solid rotor with blades inserted directly in it will serve also as a shaft.

Function	Means
Blade attachment	Straight sliding (Figure 2.1 a–d) With additional parts (Figure 2.1 e–f) Circumferential sliding (Figure 2.1 g) Arcuate dovetail (Figure 2.1 h)
Resisting centrifugal loads	Disc Drum Solid rotor
Transverse stiffness	Separate shaft Virtual shaft Drum Solid rotor

Table 2.2   Table of options for axial flow compressor rotor, based on functional analysis.

We can now draw up the table of options shown in Table 2.2. For simplicity, the blade attachments have been restricted to four categories. If every option for every function were compatible with every option for every other function, there would be

$$4 \times 3 \times 4 = 48$$

possible combinations. However, functions (3) and (5) lumped together give only four options, solid rotor, drum, discs and separate shaft, and discs and virtual shaft, reducing the total number of combinations to 16.

In the general case we can get no further. All the options are likely to be viable in certain circumstances: what is required next is some delimitation according to circumstances, and to achieve this we shall use two parameters, one measuring the difficulty of functions (1) and (3), and the other measuring the difficulty of function (5).

2.4   Parametric mapping of viable options

If r is a typical radius, say, a radius to the root of a rotor blade about half-way through the compressor, and ω is the angular velocity of the rotor, $\omega^2 r$ may be taken as a typical centripetal acceleration and $\rho\omega^2 r^2$ as a measure of centrifugal stress level (see Section 3.11). If f is a typical allowable stress, perhaps as in Figure 3.10, the ratio

$$P_{13} = \frac{\rho\omega^2 r^2}{2f} \tag{2.1}$$

is a measure of the relative difficulty of providing adequate strength against centrifugal forces; the reason for the 2 will appear in Section 3.11.

The difficulty of providing adequate transverse stiffness will depend on ρ, a typical density, and E, the Young's modulus of the shaft, virtual shaft or drum material. They will appear in a suitable parameter in the form ρ/E or some power of ρ/E. The form preferred here is $(\rho/E)^{1/2}$, which is the inverse of the velocity of sound in the shaft material if it has density ρ. The complete form of the parameter is

$$P_5 = k \frac{L^2}{r^2} u \left(\frac{\rho}{E} \right)^{1/2} \tag{2.2}$$

where u is the blade speed, L is the length of the rotor between bearing centres, and k is the ratio of the required lowest natural frequency to the running speed. P_5, like P_{13}, is a dimensionless parameter; it is the product of a factor k, the ratio L^2/r^2, which is a measure of the slenderness of the rotor and clearly makes the stiffness requirement harder to meet as it increases, and the ratio of the blade speed to the speed of sound in the material. The whole thing is a measure of the difficulty of the function (5) requirement, an incomplete measure since it does not allow for certain other variables like hub/tip ratio, but good enough.

We can now plot against our two parameters P_{13} and P_5 the areas in which the various options are likely to be suitable, as has been done in Figures 2.2–2.5. The term 'solid rotor' is here restricted to mean rotors machined from the solid but with a solidity of about one half or more.

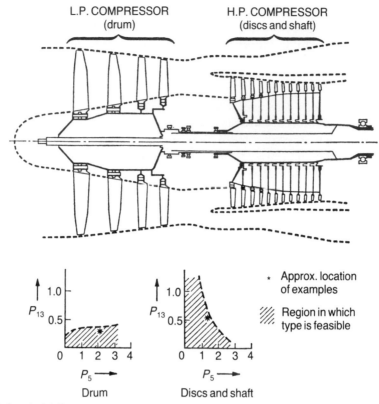

Figure 2.2   Axial flow compressor rotors: Rolls-Royce Spey. Drum construction (L.P.) and discs-and-shaft construction (H.P.) (By kind permission of Rolls-Royce Ltd.)

Drum (Figure 2.2)

The drum solution is impossible above $P_{13} = 0.5$ for then a ring of radius r is just able to sustain its own weight without the stress exceeding f (see Section 3.11). Drum solutions will scarcely be viable above $P_{13} = 0.25$. On the other hand, they are excellent for transverse stiffness and will go to as high values of P_5 as are possible.

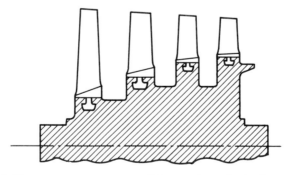

Figure 2.3   Axial flow compressor rotor, solid rotor (hypothetical).

Solid rotor (Figure 2.3)

Solid rotors are moderately good for high P_{13}, if they are fairly deeply cut away between blade rows. They can also be fairly good for transverse stiffness, if they are not deeply cut.

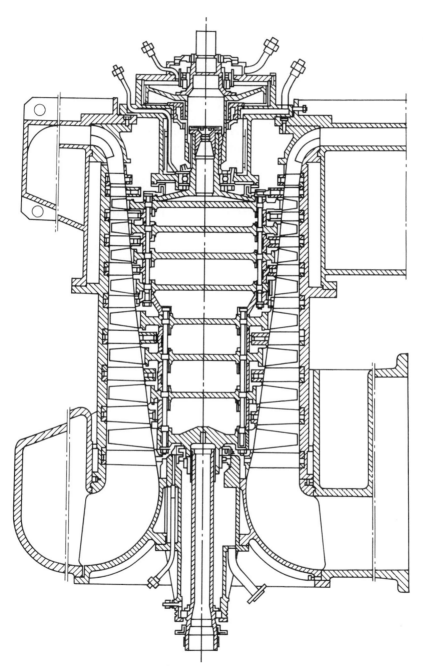

Figure 2.4   Axial flow compressor rotor, discs-and-virtual-shaft construction. (Based, with kind permission, on a figure from *Thermal Engineering*; a translation of the Russian journal *Teploenergetika* 1966 **13**, 5.)

Discs and separate shaft (Figure 2.2)

The disc design can be virtually ideal (see Section 3.11) so that the highest values of P_{13} are attainable. As the shaft radius is necessarily fairly small compared with r, only moderate values of P_5 can be reached.

Discs and virtual shaft (Figure 2.4)

When the virtual shaft diameter is small compared with $2r$, there is little difference from the separate shaft construction. It is possible, however, to put the virtual shaft at a large radius without impairing the disc strength, since the disc web does not have to be pierced.

The viable region of the virtual-shaft construction thus stretches over the viable regions of both drum and separate shaft and a good area besides in the angle between them.

We have now some idea of the regions of the 'parameter-space' P_{13}, P_5, in which different constructions are viable. We could go much further, e.g. to examine the effect of size, measured perhaps by the radius r, that makes constructions like that of Figure 2.5 possible for very large compressors. We could consider the effect of the alternative forms on cost of manufacture and weight, which would help to show which were more suitable for aircraft and land use. But enough has been done to show the general picture.

A hybrid design

One last point is this. The classification used here becomes a little woolly when designs for a high P_5 and a high P_{13} (say, around 3 and 0.5) are concerned. The only suitable constructions may be regarded as discs-and-virtual shaft with a very large diameter virtual shaft, or else as *hybrids of discs and drum*, with just sufficient drum to provide the necessary transverse stiffness (Figure 2.6). The drum part may be able to carry a little of the centrifugal load, but most of it is carried by the disc webs. The design process here may be regarded as drawing the independent optimum structures to carry out functions (3) and (5), and then marrying them together in a manufacturable whole; with as little departure from ideality as may be. The first part of this procedure can be entirely formalised ('exhausted') (see Sections 3.11 and 4.4). It is to be expected that more and more engineering products will lend themselves to such approaches as the technological targets we set ourselves become more exacting.

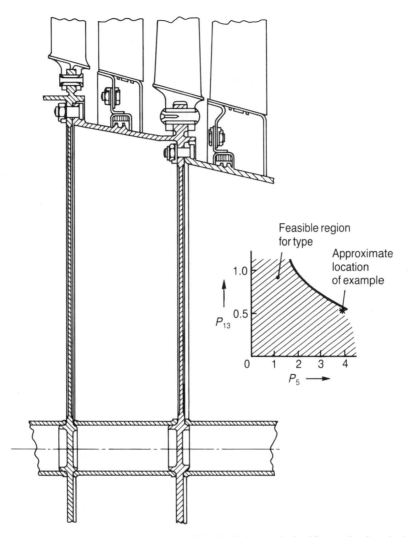

Figure 2.5   Axial flow compressor rotor with bolted virtual shaft. (Gyron Junior: by kind permission of Rolls-Royce Ltd.)

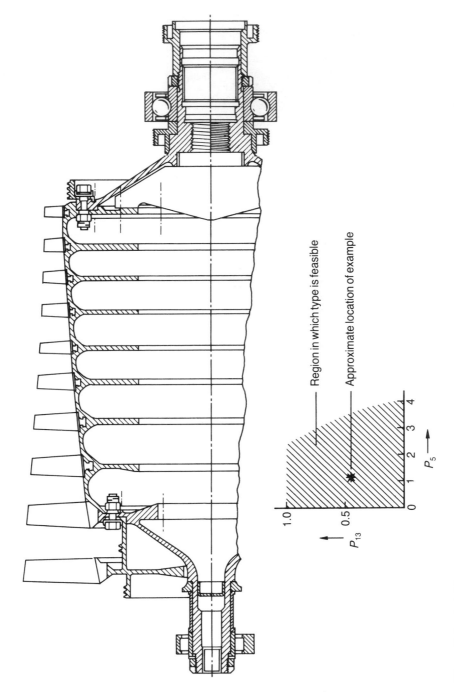

Figure 2.6　Axial flow compressor rotor, one-piece discs-and-virtual-shaft. (Gnome: by kind permission of Rolls-Royce Ltd.)

2.5   Liquid natural gas tanker: alternative configurations

This example concerns the special tankers used to transport liquefied natural gas (LNG) at substantially atmospheric pressure and about − 161 °C (112K). The tanks are thermally insulated and the heat which leaks in boils off the natural gas at an acceptably low rate, the vapour being used as part of the fuel.

A trivial functional analysis of the tank section of such a tanker shows the following principal requirements:

Containment of LNG	C (i.e. preventing leakage)
Thermal insulation of LNG	I
Structural resistance to pressure of LNG	S
Normal hull function	H

These four functions can be visualised as residing in up to four layers from inside to out. We can use a convenient code for the possible configurations, e.g.

C(IS) H

means that a layer providing containment C is innermost, followed by a single layer combining the functions of insulation I and structure S, and so on. There are in all 75 such arrangements.

Codes of this sort are useful, especially for concise and unambiguous communication between designers and for recording the results of design work. Figure 2.7 shows a possible construction corresponding to the code (CS)IH.

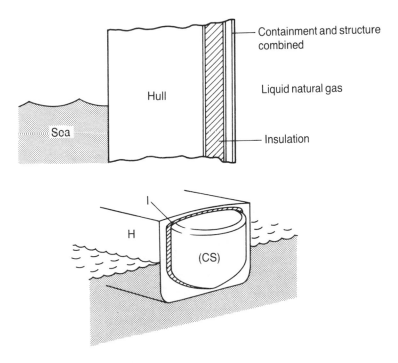

Figure 2.7   LNG tank, code (CS)IH.

Reducing the field

A few simple considerations enable us to reduce the number of alternatives substantially. For example, it is reasonable to reject all those configurations in which one layer combines other functions with I: the nature of suitable insulating materials is such as to make such combinations impracticable. This 'round of elimination' leaves only 44 of the original 75 configurations. If the I layer is outside the layer supplying the H function, the hull will be at roughly $-160\,°C$, which has among its other overwhelming disadvantages this: the hull will have to be made in a material free from brittle fracture down to that temperature, and the cheapest of such materials is much dearer than conventional shipbuilding steels, apart from increased fabrication costs. It is also difficult to imagine either C or S lying outside H, so that all orders which do not have H as one function of the outer layer can be rejected: this leaves 13 survivors.

Another elimination may be made on these lines: the two structural functions, S and H, involve resisting internal and external pressures respectively. If the two are combined in one structure, the structural tasks (see Section 8.1) will partly cancel each other and the material required for both functions will be less than for H alone. There is such a large saving here that no scheme which does not take advantage of it where possible can be regarded as viable. The only reason this *combination of functions* may not be possible in some cases is that I separates S and H, so that one is in a cold region and the other at ambient temperature, and a problem of differential expansion prevents the two structures being made to co-operate. If we simply lined a hold with insulation and then further lined the insulation with a thin metal tank resting against it, the cold would contract the metal away from the insulation, the pressure of liquid gas would try to force it back, and it would burst or tear. We therefore reject schemes in which S and H are not combined although not separated by I, leaving eight schemes.

The survivors are

CSIH*	CI(HS)	(CS)IH
SCIH	IC(HS)	I(CHS)
SICH*	SI(CH)*	

The three shown with asterisks can be eliminated on the following grounds: if C is outside I, there is no good reason why S is not also, so that S can be combined with H giving the saving mentioned above; the order CSIH can only be economical if S is a structure working principally as a membrane (i.e. *not* a plate or grid) and it is then difficult to see why it does not also perform C.

Designs used in practice

Four of the five remaining orders have been used in ships. The early British Methane Princess and Methane Progress are partly (SC)IH and partly SCIH, having polygonal free standing tanks with internal staying, and later ships have been either of this form, or pure (SC)IH or of the 'membrane' tank form CI(HS), with a thin non-structural metal tank resting on the insulation and with the structural load borne by the hull. In membrane tanks, the differential expansion problem can be overcome either by using a metal, like invar, which hardly contracts on cooling, or by using a corrugated sheet which can be stretched without developing high stresses and so will expand by flexure of the corrugations to make up for its contraction on cooling. But the first shipping of LNG on the Mississippi was done in barges lined with wooden insulation, which were I(CHS) and would have worked well except for the deterioration of the insulation.

There are thus two principal philosophies of design in LNG tankers. The first is to have a strong tank resisting the LNG pressure independently, without support from the hull, and the second is to have a non-contracting or stretchable membrane tank, leaving all the structural function to the hull. The first needs more structural strength, some in expensive metals, but the membrane construction is expensive in itself, so there is little to choose between them in cost.

2.6   Further examples of combinative treatments

Examples have been given of combinative treatments based on the analysis of function and on configurations or orders. Figure 2.8 shows various configurations of reduction gear for turbine-driven tankers, together with comparative costs and weights [5]. This is a good example of a study which has been carried through for all the alternatives in a fair amount of detail.

Examples of a classification being used as a base are uncommon. This method is only available when there are already many known designs in a field, as was the case of the Wankel rotary engine, which will be known to most readers. There have been numerous inventions of 'chamber mechanisms', i.e. mechanisms in which a chamber formed between two or more parts cyclically increases and decreases in volume. Subject to certain practical limitations, such mechanisms can be used as bases of expansion engines, i.e. steam engines, internal-combustion engines, air motors etc., in place of the familiar piston and cylinder arrangement. Wankel formed a systematic classification for all such inventions using only rotating components, with about 800 different combinations of options or pigeon holes.

Now in such a classification, some of the practical considerations, like the compression ratio that can be achieved, difficulties of sealing and providing valves, inherent balance or lack of balance and so forth, are associated with the pigeon holes rather than the particular designs [6]. Wankel devised numerous new chamber mechanisms, concentrating on those that occupied the most promising pigeon holes, and one of these has been the basis of several successful designs.

Figure 2.9 illustrates the Systematic Design Method of Rodenacker, as applied to the design of a machine used in the quality control of foil production. It punches small discs from a roll of foil according to a predetermined pattern, weighs them automatically and transmits the data for processing. A part of the method is a combinative study in which both functions and configurations are treated and Figure 2.9 shows the selected optimum means for the foil sampling machine. Figure 2.10 shows a trial assembly of these optimum aspects into a first rough scheme.

Rodenacker was one of the first advocates of systematic design. After him have come others, so that most of the German engineering schools have their exponents, and there are VDI Richtlinien which cover the subject[48]. A typical and excellent work is available in English translation (*Engineering Design*[43]) and should be read by anyone interested in functional design. However, all the more complicated systematic approaches seem to present no advantages, and some disadvantages compared with the simple table of options presented here. It is not difficult to

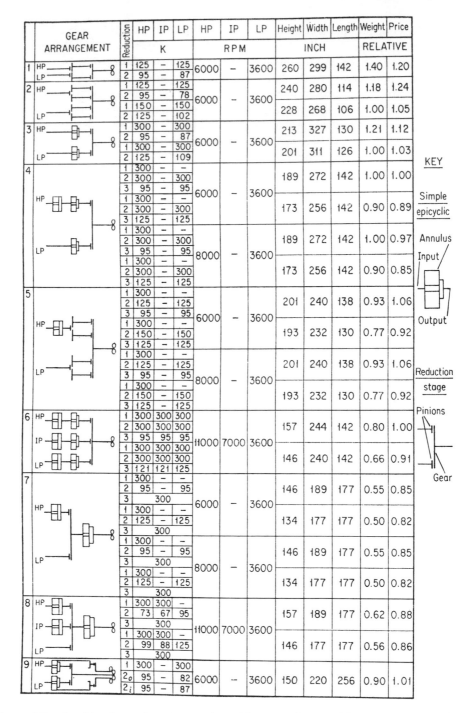

#	GEAR ARRANGEMENT	Reduction	HP (K)	IP (K)	LP (K)	HP (RPM)	IP (RPM)	LP (RPM)	Height (INCH)	Width (INCH)	Length (INCH)	Weight (REL.)	Price (REL.)
1	HP / LP	1	125	–	125	6000	–	3600	260	299	142	1.40	1.20
		2	95	–	87								
2	HP / LP	1	125	–	125	6000	–	3600	240	280	114	1.18	1.24
		2	95	–	78								
		1	150	–	150				228	268	106	1.00	1.05
		2	125	–	102								
3	HP / LP	1	300	–	300	6000	–	3600	213	327	130	1.21	1.12
		2	95	–	87								
		1	300	–	300				201	311	126	1.00	1.03
		2	125	–	109								
4	HP / LP	1	300	–	–	6000	–	3600	189	272	142	1.00	1.00
		2	300	–	300								
		3	95	–	95								
		1	300	–	–				173	256	142	0.90	0.89
		2	300	–	300								
		3	125	–	125								
		1	300	–	–	8000	–	3600	189	272	142	1.00	0.97
		2	300	–	300								
		3	95	–	95								
		1	300	–	–				173	256	142	0.90	0.85
		2	300	–	300								
		3	125	–	125								
5	HP / LP	1	300	–	–	6000	–	3600	201	240	138	0.93	1.06
		2	125	–	125								
		3	95	–	95								
		1	300	–	–				193	232	130	0.77	0.92
		2	150	–	150								
		3	125	–	125								
		1	300	–	–	8000	–	3600	201	240	138	0.93	1.06
		2	125	–	125								
		3	95	–	95								
		1	300	–	–				193	232	130	0.77	0.92
		2	150	–	150								
		3	125	–	125								
6	HP / IP / LP	1	300	300	300	11000	7000	3600	157	244	142	0.80	1.00
		2	300	300	300								
		3	95	95	95								
		1	300	300	300				146	240	142	0.66	0.91
		2	300	300	300								
		3	121	121	125								
7	HP / LP	1	300	–	–	6000	–	3600	146	189	177	0.55	0.85
		2	95	–	95								
		3	300										
		1	300	–	–				134	177	177	0.50	0.82
		2	125	–	125								
		3	300										
		1	300	–	–	8000	–	3600	146	189	177	0.55	0.85
		2	95	–	95								
		3	300										
		1	300	–	–				134	177	177	0.50	0.82
		2	125	–	125								
		3	300										
8	HP / IP / LP	1	300	300	–	11000	7000	3600	157	189	177	0.62	0.88
		2	73	67	95								
		3	300										
		1	300	300	–				146	177	177	0.56	0.86
		2	99	88	125								
		3	300										
9	HP / LP	1	300	–	300	6000	–	3600	150	220	256	0.90	1.01
		2o	95	–	82								
		2i	95	–	87								

KEY

Simple epicyclic
Annulus Input
Output
Reduction stage
Pinions
Gear

Figure 2.8 Combinations of marine gearing with two or three turbine shafts: results of a comparative design study. The diagrams in the left-hand column indicate the configuration under study, e.g. the third diagram down shows the turbine divided into two sections, HP and LP, each driving into a simple epicyclic. The output shaft of each epicyclic drives a pinion meshing with a single large gear on the propeller shaft. The last diagram shows a version for contra-rotating propellers, in which drives are taken from both the output shafts and the annulus gears of the epicyclics (see section 8.3). Studies have been made of the effects of three degrees of freedom—division of the turbine into two or three sections (see

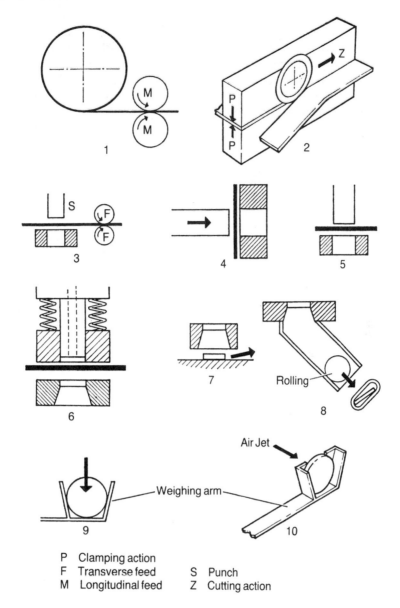

P   Clamping action
F   Transverse feed   S   Punch
M   Longitudinal feed   Z   Cutting action

Figure 2.9   Systematic Design Method of Rodenacker (1). The example shown is a machine for automatically recording statistical data on the distribution of mass per unit area in a sample roll of foil, for quality-control purposes. The data are obtained by punching out of the foil and individually weighing a pattern of small discs. The diagrams illustrate the separately optimised means of performing the functions 1–10, where 1 is feeding lengthways, 2 is cutting in widthwise strips, 3 is feeding the strips sideways to the punch which punches out the discs (functions 4, 5, 6) etc.

Figure 2.8 *continued*
sections 5.3–4) different materials (shown by the different K's) and different HP turbine speeds—besides the different gear arrangements. The factors K are a measure of tooth stress, i.e. they are proportional to the constant C in equation 4.13. (By kind permission of Dr I. Jung and the Society of Naval Architects and Marine Engineers.)

produce extended lists of separate functions and means of solving these separate functions, or tables of options, and it is a useful procedure. But the difficulties arise in the reduction of such tables, and here the writer believes the suggestions made in this chapter are helpful and apt to the use both of students and practised engineering designers. Also, additional bases besides function, such as configuration, can often be used with advantage.

2.7   Elimination procedures for tables of options

A table of options constructed in one of the ways suggested in Section 2.2 may be regarded as a large number of schemes in embryo. In some cases, such as the tanker gearing shown in Figure 2.8, it may be practicable to study all the feasible combinations reasonably thoroughly, but in the more typical case, like the LNG tanker, only a few variations can be studied in depth, and most of them must be eliminated by methods of a more cursory nature. Here it is convenient to consider the tables based on functional analysis separately from those based on order.

Occasionally it may be possible to decide on the best means of performing each function separately, and by combining these options obtain the best combination, i.e. the optimum solution is also the optimum in respect of each function considered independently. More often, however, the best means selected in this way will be incompatible, or more expensive to combine, so that the best overall solution will incorporate some other alternatives.

In the foil sampling machine (Figures 2.9 and 2.10), it proved possible to optimise each means separately. This is often the case with plant that carries out a series of processes, where the number of kinds of interaction between one function and the ones that precede and follow it are limited. We shall describe such design situations as of weak or limited interaction.

As a simple example of strong and complex interaction, consider the various advantages listed in Table 2.1. The first three are associated with particular options r, F and f respectively. But the fourth advantage, no propeller shaft and a clear floor space, is not associated with a particular option—indeed, the two combinations, Ff and Rr, which share it have no common option. In the language of the statistician, this particular advantage is a *first-order interaction*. In the same way, a *second-order interaction* would be an advantage, disadvantage or other consequence associated with three options taken together. It is convenient to use the term 'interaction' both in this narrow and precise sense and also in the vaguer way where it describes only the general order of interconnectedness of problems.

Evaluation of options

One way of finding the best combination is to put *values* on each option and on each interaction and find the total values for all the possible combinations. Indeed, by putting a very large negative value on the interaction of two options which are incompatible, we can reduce the problem to one of finding the highest total value, regardless of whether combinations are possible or not.

Thus, in the case of the motor-car configuration problem, we can consider the values of the two alternative locations of the engine, F and R. The only aspect

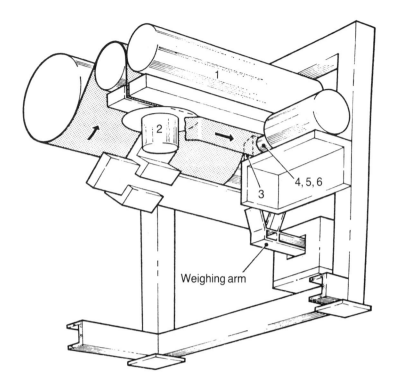

Figure 2.10   Design Method of Rodenacker (2). Here the selected means of Figure 2.9 are shown assembled into a first schematic design (the same numbers are used for parts which are visible). Note that either or both of the two motions of functions 1 and 3 could have been given to the punch instead of the foil —*cf.* the IBM typewriter of section 2.13.

allowed for in this simple treatment is the effect on road-holding, so that we might put $F = 2$, $R = 0$, say. The 'zero position' is of no consequence, so we have taken R as zero quite arbitrarily.

With regard to the drive location, the front-wheel drive is advantageous for road-holding but involves the driving shaft flexing through the large angle needed for steering. Suppose for the moment the disadvantage outweighs the advantage and put $f = -1$, $r = 0$.

Finally, we need to consider the interactions. In this very simple problem, there are only two choices to be made, where to put the engine and the drive—what will be called here 'two degrees of freedom of choice': consequently the first-order interaction is the only interaction. We assume that the only important consideration is the disadvantage of a propeller shaft, rated at -2, so that for the interaction we put

$$\overline{Ff} = \overline{Rr} = 0, \overline{Fr} = \overline{Rf} = -2$$

where the bar is used to distinguish the *interaction* Ff from the *combination* Ff.

The total values for the combination are then:

$$Ff = F + f + \overline{Ff} = 2 + (-1) + 0 = 1$$
$$Fr = 2 + 0 + (-2) = 0$$
$$Rf = 0 + (-1) + (-2) = -3$$
$$Rr = 0 + 0 + 0 = 0$$

This reflects the real position for small cars today crudely but fairly realistically. In the past, the difficulty and consequent expense in providing for large flexure in a front-wheel drive favoured Fr and Rr relatively to Ff, but the introduction of reliable cheap high-angle joints to the engineering repertoire has changed the situation.

What has been done here with a very simple problem can clearly be done in much more complicated cases without great difficulty. However, it is easy to be led astray by elaborate evaluation schemes of this sort, which can lead to absurd conclusions. Typically, a method will be used in which candidate schemes are assessed in respect of a number of different aspects and given marks on each. These aspects are given different weightings, perhaps on a scale of one to ten, and a single mark produced for each scheme by multiplying its marks for each aspect by the appropriate weighting, and then adding all these weighted marks. Both the marks and the weighting are generally highly subjective, and the rigidity of the process and often also the size of the 'matrices' considered, hinders the proper operation of judgement.

A much more reliable and reasonably expeditious approach is to compare one scheme with another, as is explained in Section 2.8(c).

2.8   Other ways of reduction

In Section 2.7 it was shown how elimination may be performed on tables of options which are particularly amenable to formal treatment. In Section 2.3 it was shown how, in other cases, by quantifying certain aspects of the design problem, parametric plots may be produced that help in the selection of the best combination of options. In most applications, however, there is a need for methods of greater flexibility (and hence usually less rigour), which with small expenditure of effort can narrow down the field of enquiry without much risk of missing a good solution.

Such methods must be often *ad hoc*, devised piecemeal to suit the task in hand, and resting largely on engineering judgement and common sense. Nevertheless, a number of ideas are given in the following notes: in most problems several of these will be of assistance.

(a) Grouping of lines of the table of options

Each choice between options corresponds to a line in a table of options written in the form of Table 2.2, which has three lines corresponding to functions 1, 3 and 5, respectively. It is useful in more elaborate cases to order these lines vertically, so that they fall into groups with most interactions confined within groups and as few and weak interactions as possible between members of different groups. The mathematical problem involved has been treated by Alexander [8], but a good appoach to the ideal restructuring of the table can be made by trial-and-error: the natural structure of the problem is often a great help here, as the groups tend to be related to functional or spatial groups in the real device or system.

(b) Amalgamation or condensation of lines

If two lines in a table consist of m and n options, respectively, they can be amalgamated or condensed into a line consisting of the mn combinations they form. This is no improvement unless, as often happens, a substantial number of these *mn* combinations are impracticable or meaningless. In the axial flow compressor rotor example, the second and third lines of Table 2.2 have three and four options, but they amalgamate into a line with only four options.

A very satisfactory way for elimination to develop is for several groups to reduce in this way to single lines with only a few options each. To achieve this it is necessary to reject combinations and recognise incompatibilities not only of an absolute kind, but also where quick rough calculations and engineering judgement indicate *prima facie* economic disadvantages too large for there to be much chance of their disappearing on more careful appraisal. One example of this approach was the rejection of orders with S and H adjacent but not combined in the LNG tanker study (Section 2.5). An important safeguard in such cases is that the designer should return and reconsider his verdict later, when his insight into the problem has improved.

(c) Elimination of single options

Sometimes a short study will suffice to show that one option is inherently inferior to another, whatever the combination in which it is used, and then a slight reduction in complexity can be made by rejecting the inferior option. In the study of disc-like structures in Section 3.11, it is shown that spokes or a ring are inferior to a disc as regards economical use of material in resisting large radial loads, by a factor of at least two. In the case of the ring, this is offset by the fact that this structure, extended axially to become a drum, is very good for resisting transverse bending. Consequently a ring or drum design may be viable where the radial strength requirement is not very high and the transverse stiffness needed is large, as was seen in Section 2.4. However, the spoked design would appear to have no compensating advantage and might well be rejected, even without considering the severe load-diffusion problems at the hub.

A good practice in such cases is to try and imagine some circumstances in which the option about to be discarded might have some advantage. For example, the spoked wheel might have some virtue for a compressor of very low solidity, say, a very light machine for compressing a gas at a fraction of an atmosphere, were such a thing ever to be needed (to be convinced of this, study the spoked bicycle wheel and try to find a viable alternative). Another application is to a design using a high-strength filament such as carbon, which is most economically used under uniaxial stressing. Here, no doubt, the outward appearance would be that of a disc, but the true nature might well be spoked.

This is about as much as can usefully be said about elimination without bringing in elaborate notations that are more trouble than they are worth, or giving extended examples. Every user will soon develop his own approach, but a general suggestion may be of use here; it is most economical of effort to be brutal and sweeping to begin with, to arrive quickly at a small group of preferred basic schemes, and later to revise the elimination process in a charitable and open frame of mind, to decide whether some of the eliminated ideas are not worthy of reinstatement; this latter process the writer calls 'repêchage', and it will be referred to again (Section 2.13).

However elimination is tackled, two characteristic difficult comparisons arise. The first is between very like solutions and is therefore generally not of the utmost importance: *differential studies* (see Section 4.3) are often a help here. The second is between very different solutions, and in the absence of fundamental methods of comparison fairly detailed and expensive design studies may be necessary to resolve the matter. Indeed, if the sum eventually to be spent on hardware is small enough, it will be cheapest to leave the question unresolved and 'plump' for one or the other.

Where the problems of straight elimination are overwhelming, the techniques of the next section can be used.

2.9   Synthesis from tables of options: kernel tables

We have already seen, in the examples of the axial flow compressor rotor and the LNG tanker, how fuller insight into a problem, treated first by combinative methods, leads to a principle for synthesising a good design. This state of affairs is to be expected when the technical problems are especially difficult, e.g. the synthesising principle only becomes apparent and of value when the compressor rotor requires great radial strength combined with high transverse stiffness. It is commonly based upon some quantifiable and readily costed task to be performed by the design, as in the LNG tanker where the structural tasks provide the starting point.

Even when no natural principle of synthesis emerges, we can tackle the problem of selecting preferred solutions from a table of options, not by rejecting the bad, but by building up only the good. In the extreme case, we take some particular line in the table of options and one particular option from that line. We then choose another option from a second line, on the basis of it being the best to combine with the first or *pivotal option*. We then take a third line, and the conditional best option from this line, the conditions being our previous choices from the first and second line, and so on. This process will be called synthesis from a pivotal option. Clearly, its success depends on the choice of the first line and the order in which successive lines are taken; it needs much judgement and is difficult to apply 'from cold', before insight into the particular problem has been developed.

An alternative procedure that combines analysis and synthesis is to construct a small table of options out of a handful of the most important lines in the original table. This small or kernel table is then reduced to a few preferred incomplete combinations by the sort of elimination process already discussed. We now proceed by synthesis as with the pivotal option, but taking one of these preferred incomplete combinations from the kernel table as a pivotal set of options. This is done with each of the preferred incomplete combinations from the kernel table, giving a set of preferred complete combinations. Thus, in the car example of Section 2.1 we could regard Table 2.1 as a kernel table, in which Ff and Rr might be our preferred incomplete combinations. We might then work through a larger table including such choices as form of engine, method of cooling, orientation of crankshaft, types of suspension, so as to end with two schemes, one based on Ff and one on Rr. It is likely, however, that some important choices will be found where the decision on Ff, say, will not make selection any easier, two options, a and b say, seeming to have equal merits; we then increase the number of preferred schemes, pursuing the process with Ffa, Ffb, and Rr.

These synthesising approaches, where a few good combinations are built up rather than vast numbers of bad combinations being eliminated, seem much more natural and appropriate to the designer. Perhaps the best way is to start with a table of options, to work a little with it by elimination, rejecting altogether inferior options, reordering lines and amalgamating lines, and then to turn to synthesis, based on either a pivotal option or, more probably, a very small kernel table of two or three lines. (The largest problem of this kind that the writer has tackled needed a five-line kernel table, but only five preferred incomplete combinations came out of it.) With luck and good management there will emerge a handful of preferred complete combinations, in two or three 'families', with large differences between families and only small differences within them.

It is to be hoped, also, that each family of preferred complete combinations will have its own design philosophy or rationale or synthesising principle, which the designer or designers will by now understand fairly well. If so, this is the appropriate stage to introduce considerations that are foreign to the original table of options and which would have overburdened it had they been put in at the start. These matters are discussed in the next section.

2.10   Evolutionary techniques: hybrids

The combinative techniques here described are termed 'evolutionary' because they produce new schemes from combinative treatments that do not explicitly contain them. For example, some promising solutions to the LNG tanker problem are based on hybrids, mostly CI(SH) but partly (CS)IH (see Section 2.5). Such hybrids are not allowed for in the original permutative scheme, but arise naturally from a conscious synthesising principle or design philosophy, which in turn grows out of deeper understanding of the problem and the more abstract approach that this understanding makes possible. The processes that produce these hybrids are much more economical than the 'blind' combinative methods since they produce only promising ideas. On the other hand, they are not systematic and must be developed specially for each new problem.

To explain this by an example, let us take the synthesising principle behind the hybrid CI(SH)–(CS)IH tank. The ideal, already explained in Section 2.5, is a thin-walled tank, C, which merely contains the LNG and transmits the pressure load via the insulation, I, against which it rests to the hull (SH)—this would be a purebred CI(SH) design. The trouble is that the space occupied by the tank or lining, C, is governed in size by the hull, which remains warm and does not contract when the tank is filled with liquid LNG at about −160°C. If we are to keep a pure CI(SH) design, the lining, C, must not contract at this temperature, either because it has a very low coefficient of expansion and so can withstand being fixed at the corners, or because it is cunningly corrugated. But suppose the tank does contract but is also deformed in some way; since its superficial area is reduced, it cannot still fit the hold everywhere, but it might fit it *nearly* everywhere. Where it does not fit, it will have to provide structural strength of its own to withstand the cargo pressure, i.e. it will have to be (CS)IH or possibly SCIH (it turns out that we may need 3–30 per cent of self-supporting wall, according to materials, type of design, and the size and complexity of tank).

If the tank is basically a polyhedron, the obvious type of deformation is a bulging of the flat faces. Separation takes place at the edges, and they must be made self-supporting or at any rate supported otherwise than by the insulation. Also the 'bulging stresses' must somehow be kept within acceptable limits.

In taking this problem so far, two fairly abstract ideas have been used. The first was the idea of costs associated with quantifiable structural *tasks*, which led us to a strong *a priori* preference for solutions with S and H, the two structural functions, combined in one 'layer'. The second was the hybridising idea outlined above. Both lead to particular arrangements being adopted—they are synthesising principles— but the second is narrower and more powerful. By an evolutionary process a design philosophy has grown out of the original combinative treatment.

Note that a common starting point for an evolutionary step, often of the hybridising sort, is an attempt to produce a scheme to combine the advantages of two other schemes; alternatively, the object may be to avoid both the different weaknesses of its two parents, or to eliminate a single disadvantage from an otherwise promising scheme.

At this point it is convenient to study one of the richest design problems which has ever been tackled, where a great variety of solutions have been studied and where many of the aspects so far dealt with have appeared—different means of performing functions, different configurations and hybridisation.

2.11   An example: wave energy converters

Men have long seen sea-waves as a potential source of energy, and in the mid 70s the impending energy crisis, brought to public notice by oil price rises, led to considerable study of wave energy converters (WECs) mostly in the UK but also in Japan, Scandinavia and North America. This activity has since subsided because of unfavourable costs and the receding of the energy crisis, but WECs provide a very interesting design problem, particularly because of the large number of possibilities—the subject is an inventor's paradise.

Before analysing functions and making a table of options, it is helpful to look at a simple example, and one of the oldest designs will suit: a 'bobbing buoy'.

Imagine a buoy moored to the seabed by a taut cable containing in its length a hydraulic cylinder (Figure 2.11). When a wave crest passes, immersing the buoy to a greater depth, the buoyancy, and hence the force in the cable, will increase. When a trough passes, the force in the cable will be reduced. This fluctuating force can be made to work the cylinder as a pump, extracting energy from the sea. This example enables us to identify the principal functions in a wave energy converter.

First, there must be a working surface (as Rodenacker would say) where the waves act on the device; in the example, it is the surface of the buoy. Secondly, there must be some means of providing a reaction against the wave force, for if it is unresisted the wave can do no work; in the case of the buoy this reaction comes via the cable from the seabed. Finally, there must be the work-extracting device, here represented by the hydraulic cylinder-cum-pump. A much more complicated analysis of function is possible, but is not necessary here.

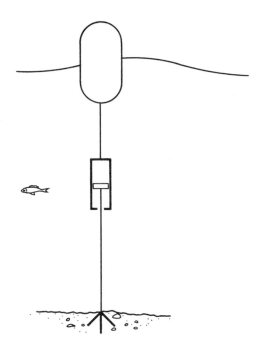

Figure 2.11   Simple wave energy converter.

Table of options for a WEC

We can now make a table of options. The working surface can be a solid, a liquid or a gas, and, if it is a solid, it can be either rigid or flexible (there is no sharp distinction in theory between rigid and flexible, but it is plain enough in practice). For providing the reaction to the wave force, there are three options: fix the device to the seabed ('fixing'), balance one wave force against another ('balancing') or use inertia ('inertia'). For power extraction, the general engineering means available are pneumatic, oil hydraulic, water hydraulic or direct electrical, and the fluid machines can be of displacement or of kinetic type. These aspects are not very important in the early stages and will be omitted. What does complicate the issue, however, is the aspect of configuration.

The simplest configuration of device we can imagine is the 'terminator', an elongated machine lying parallel to the wave crests, rather like a breakwater, at which the waves are absorbed and their energy converted into some useful form. The buoy is an example of another configuration, the point absorber, which draws in energy to itself from the passing seas. Finally, there is the attenuator, an elongated form lying at right angles to the crests of the waves, which run down its sides losing energy as they go. It may help to fix these ideas to think of aerials for television sets, which are also devices for gathering in energy from a sea of waves (in this case, electromagnetic waves), the object being the information the energy carries, not the energy itself. The point absorber corresponds to the single dipole aerial and the attenuator to the linear array of dipoles pointed towards the transmitter.

We can now form our simple table of options, leaving out liquid in the first line because it seems unlikely to have any practical embodiment.

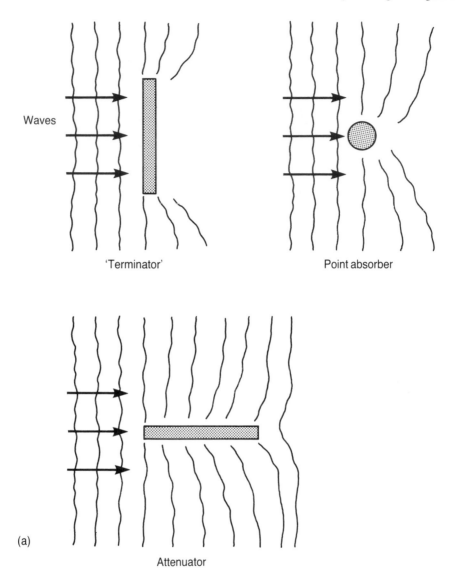

Waves

'Terminator'   Point absorber

(a)

Attenuator

Figure 2.12a   Plan views of WEC configurations and wave patterns.

Working surface Rigid solid (S_R) flexible solid (S_F) gas (G)
Reaction means   Fixing (F) balancing (B) inertia (I)
Configuration   Terminator (T) attenuator (A) point absorber (P)

On this basis the bobbing buoy, with a rigid working surface, reacting against a fixture in the seabed and with a point absorber configuration is

$$S_R FP$$

Nearly all the 27 combinations have been studied in some depth.

Many designs of WEC are based on an oscillating water-column (Figure 2.12b) in which the rise and fall of the water in a vertical tube forces air both ways through a Wells turbine (designed to accept flow in either direction) or, by means of valves, one way through a conventional turbine. In this case the water acts directly on a

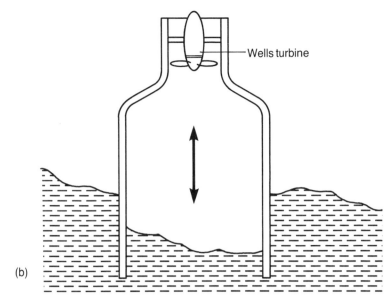

(b)

Figure 2.12b   Simple oscillating water column (OWC).

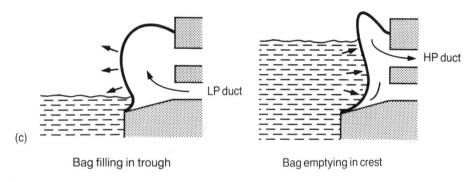

(c)

Bag filling in trough   Bag emptying in crest

Figure 2.12c   Lancaster flexible bag.

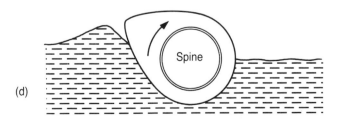

(d)

Figure 2.12d   Salter's 'duck'.

working surface of air (G). In others, the wave crests squash air bags which act as bellows, forcing air through turbines: the bags later refill in the wave troughs (Figure 2.12c). Such a device is S_F. In the Salter Duck, a cam-like body turns under wave action on a central cylindrical shaft, or spine (Figure 2.12d), so this device is S_R.

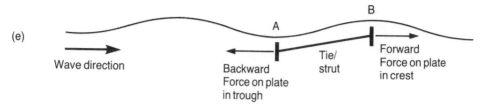

Figure 2.12e   Balancing of wave forces in a WEC: attenuator with plates.

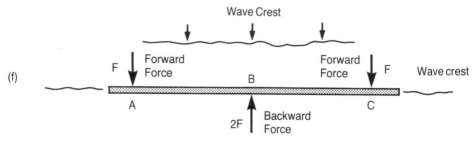

Figure 2.12f   Balancing of wave forces in a WEC: terminator.

A dominant function

In the event, one aspect tends to dominate the problem of wave energy at reasonable cost, that of providing a reaction. Originally, it seemed that both inertia and fixing to the seabed would be prohibitively expensive, leaving only the balancing of one sea force against another. However, to balance two such forces they should preferably be 180° out-of-phase. If we seek an out-of-phase force in a train of waves, we need to go half a wavelength in the direction of travel, e.g. we find a forward force in the crests and a backward force in the troughs (Figure 2.12e). This consideration should give an advantage in structural cost to an attenuator, but this is offset by the fact that the energy-gathering capacity is inferior to that of a terminator of the same length by about 25 per cent.

It should be noted that the structure ABC in Figure 2.12f, the terminator hull, is a beam. The structure AB in the attenuator is a tie/strut, but it could have been a beam, as it is in the Lancaster Flexible Bag WEC (Figure 2.12c). A balancing beam, however, has to be twice as long as the tie/strut since it has to balance a central force against two end forces 180° out of phase. But in any case, a slender beam is an inefficient structure, in that it requires several times the minimum amount of material that will perform the same structural task.

Design philosophies for WECs

These considerations lead to examples of the emergence of a design philosophy. In the case of the Lancaster Flexible Bag, the first thought was concerned with the violence of the sea, which led to a preference for a flexible working surface with the seas running along it. Moreover, it seemed better that the device should ride head to sea, rather than receiving the waves broadside on, in the 'broached to' condition feared by mariners. This led to rows of flexible bags down the sides of an attenuator, a form which was also favoured on the grounds of structural economy. A lower energy-gathering capacity was suspected, and later demonstrated, but was not felt to outweigh the advantages.

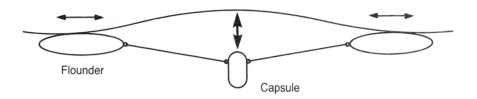

Figure 2.13   'Flounder'—a WEC based on a simple design philosophy.

An alternative philosophy is shown by the Salter Duck (Figure 2.12d) in which cam-like bodies turn on a central spine which acts structurally rather like the beam ABC in Figure 2.12f (where the forces F do not lie wholly in the horizontal plane). When Salter found that the spine would be subject to excessive bending moments, he provided it with hinged joints so that it could bend, rather than break. Bending of the joints is resisted by hydraulic cylinders which work as pumps, providing extra power extraction means in addition to the 'nodding' of the cam-like bodies or 'beaks'. Power from these is obtained by resisting hydraulically the precession of gyroscopes in the beaks under the nodding action, an ingenious method which enables all the machinery to be within a sealed chamber.

For all its ingenuity, the Salter Duck bears the signs of what is sometimes called 'Christmas tree' design, whereby each difficulty is met by adding some further complexity. While much good design necessarily goes this way, a point may be reached where the question should be asked, would it not be better to start again with a different concept? Incidentally, the Salter design with its two modes of power extraction is essentially a hybrid solution.

Yet another philosophy leads to the design shown in Figure 2.13; its essence is to start from the choice of balancing to provide the necessary reaction as a pivotal option, and to choose the cheapest possible structure, which is probably a cable pretensioned to be able to act as both tie and strut, like a bicycle wheel spoke (Section 2.8). This structure is then used in a simple attenuator, as in Figure 2.12e. It turns out to be better to use bodies of the form shown, rather than simple plates, but, although large, these are simply inflatables, filled mainly with seawater. The bodies move to and fro ('surge', technically) under wave action. The dense capsules hung from the cables tension them and provide stiffness which increases power extraction. They also execute violent vertical oscillations because of the toggle action of the cable, and contain inertial pumps or compressors which provide the power take-off, in an environment hermetically separated from the sea. In its final form, this device ('Flounder') seemed at least competitive with any other (but see Chapter 9).

Many other interesting design philosophies have been followed in WECs, notably the Shrivenham Triplate and the Bristol Cylinder (which rest on elegant bases in physical science), the Cockerell Raft (which derives from the physical insight of the inventor of the hovercraft), the Russell Rectifier (which has, as the name suggests, a close parallel in electrical engineering), and a family of oscillating water column WECs, starting with the Masuda buoy, whose working surfaces, being of air, are not susceptible to damage—a most attractive aspect.

Altogether, the design of WECs has called forth a great display of talent and ingenuity, and the subject is well worth study on that account alone.

2.12   Evolutionary techniques: redistribution of functions

Improvements in schemes can often be effected by redistributing functions among parts, for example, by using two separate parts to perform two functions that were previously performed by one part. A historic case is Watt's improvement of Newcomen's steam engine. Newcomen used the cylinder of his engine as a condenser also, spraying cold water into it at the end of each stroke. Unfortunately, this cold water not only condensed the steam but also cooled the cylinder walls, so that the first steam admitted on the next stroke was wasted in heating the cylinder again. By using a separate spray condenser alongside the cylinder, Watt reduced the fuel consumption per unit of work by about two-thirds [9].

As an example of the reverse process, of taking two parts each with its own function and combining them into a single part to perform both functions, consider the widespread substitution of a 'unit construction' car body for the earlier system of chassis and separate 'coachwork'.

There are also cases where one function is divided among several parts, as in multi-engined aircraft, and cases where one part with its own proper function helps with the function of another part, as when the structural concrete of a nuclear reactor for power generation helps also to provide the shielding.

Most cases of redistribution of function are largely matters of common sense, and detailed discussion of them would be trivial; the important point for the designer is that he should not overlook the possibilities this technique affords. However, it is probably worth examining a few of the less obvious aspects and such general indications (in the medical sense) as are available.

(1) Distribution of two functions, one to each of two parts, when one part might have done both

This is perhaps the possibility that is most often overlooked when it should not be, sometimes because the two separate functions involved are not recognised as such. In nuclear reactors with reinforced-concrete pressure vessels, the structural function resides in the prestressed steel reinforcing tendons and their concrete matrix, but the containment or sealing function is filled by a thin steel shell (Figure 2.14). This is a cheaper solution than combining the two functions in a thick steel shell, as was done earlier, so much so as to make it reasonable to put the whole gas circuit inside one pressure vessel and sweep away many knotty problems of duct design, differential expansion, and so forth.

A close parallel is the Philips hermetic piston seal shown in Figure 2.15. The seal is a 'rolling diaphragm', which works like a sock being turned inside out. Such diaphragms can be made to stand some tens of atmospheres pressure difference by reinforcing with fibres, but the number of reversals which can then be sustained is not sufficient for many applications.

Philips use an unreinforced polymer that can only stand a small pressure difference at which, however, it will survive hundreds of millions of reversals. The space outside the diaphragm is filled with oil at nearly the same pressure as that of the gas inside that is being kept hermetically sealed, so that the pressure difference over the diaphragm is small. Nearly all the pressure drop occurs over the outer oil seal which naturally leaks oil. However, oil, unlike gas, is easily recovered and pumped back. This seal is used on Philips Stirling cycle and reversed Stirling cycle

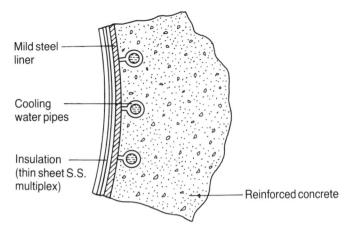

Figure 2.14   Wall of pressure vessel, gas-cooled reactor.

engines and refrigerators, where the working fluid is a small quantity of high-pressure hydrogen or helium in the cylinders which should not be allowed to escape, even slowly.

Notice the way in which the cylinder is stepped so that the volume of the oil space in the seal does not vary; this enables the elasticity of the diaphragm to nearly balance the pressure on its two sides.

The Philips C70 reversed Stirling cycle refrigerator is recommended for study as an example of refined and elegant design, combined with a proper regard to cost, as shown by the use of crankcase, crankshaft and connecting rods from an existing design of air compressor[10].

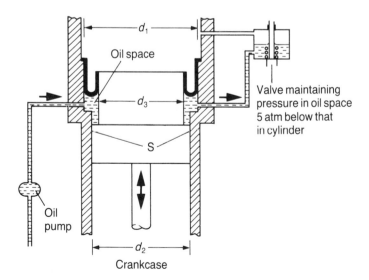

Figure 2.15   Philips rolling diaphragm hermetic seal. In this hermetic seal for pistons moving at high speeds, most of the pressure drop between the cylinder (at 100–200 atmospheres) and the crankcase occurs across the conventional seal S, only about 5 atmospheres differential pressure being carried by the rolling diaphragm. By stepping the piston diameter as shown and making the annular area between diameters d_1 and d_2 equal to that between d_2 and d_3, it is arranged that the volume of the oil space does not vary as the piston reciprocates (see section 2.12). (By kind permission of Philips NV, Eindhoven.)

Both the reactor pressure vessel and the Philips seal involve the separation of a sealing function from a pressure-resisting, that is, a structural, function, which may be conveniently thought of as 'preventing leaks' and 'preventing bursts' respectively. This particular pair of functions frequently require separating, e.g. in gear-pumps where the internal pressure tends to force the cover plates away from the ends of the gears, permitting a leak back from the high pressure at the outlet to the low pressure at the inlet. This can be overcome by providing a sealing plate between the structural cover plate and the gears. Between the sealing plate and the cover plate, sealed cavities are formed which communicate with the interior of the pump at points carefully chosen to give the right pressure in each cavity, so that the sealing plate is gently pressed against the gears, preventing leaks.

Indications

Separation of function of this kind is most likely to be of value where one or both functions are very exacting or at the limits of available means, or where the parts are very large or costly.

(2) Distribution of one function between several similar parts

Obvious reasons for adopting this technique are greater reliability, greater flexibility, e.g. in coping with part load etc., using available space, and so on. A less obvious one is to save weight, as in the case of lift engines for v.t.o.l. (vertical take-off and landing) aircraft; because of the '$1\frac{1}{2}$ power' law (see Section 4.7) there is a finite thrust for which the weight per unit of thrust is a minimum. The same law means that a number of small Pelton wheels may be cheaper than one large one (see Section 5.5).

   Another important reason for using several parts is for structural purposes, where load diffusion is concerned (see Section 7.5).

(3) Division of one function between several dissimilar parts

Very often this is the key principle in a hybrid scheme, as in the hybrid power systems used in naval vessels and military vehicles, where a low specific-weight gas turbine allied to a low fuel-consumption internal-combustion engine gives a high top speed combined with a long range at cruising speed. Another example is the use of backing pumps in conjunction with high-performance vacuum pumps.

Indications

The indications are usually pretty plain once the possibility is considered: either a wide range of requirements, some better met by one means and some by another, or a task too difficult to be met by any one means unassisted.

(4) Combination of several functions in one part: cut-price solutions

To combine functions in one part is so often a conscious part of a designer's aims that it hardly merits special mention. What is worthy of note is the aspect of what the writer calls 'cut-price' solutions. Consider the following description:

> The device consists of a store of solid fuel, means to melt this fuel at a steady rate and store it temporarily in a reservoir, and means to raise a stream of melted fuel from said reservoir into a region where it is further heated and vaporised and finally burnt in a stream of air.

This all sounds complicated, but it is simply a functional description of a candle, rather in the style of a patent. The candle itself is the store; the flame steadily melts it and forms a cup-shaped reservoir of liquid fuel round the wick. By capillary action the wick raises the liquid fuel, which is vaporised and burnt in the convection current of air drawn up the sides of the candle, which also cools the solid rim of the cup and prevents it from melting.

Here we have a device whose function is complicated, as Michael Faraday so beautifully explained, but whose physical realisation is very simple. When the description of a device has been framed in abstract terms, as here for the candle, we should always hope for and look for such a cut-price embodiment, though we shall usually be disappointed. Indeed, the abstract description of a device may be so involved compared with other ways of doing the same job that it is only worth consideration in case a 'cut-price' embodiment can be devised.

Indications

Cases where this technique is particularly indicated are those where the technical tasks are fairly easy or the volume of production large, military equipment and other cases where weight or space is at an especially high premium.

Auxiliary functions

It is often worth introducing an additional (auxiliary) function in order to make an exacting function easier to fill. For example, the pressure vessel in Figure 2.14 has an insulating lining and cooling pipes in the walls in order that the members performing the very exacting structural task do not have also to contend with high temperatures or high thermal stresses.

2.13 Repêchage

It has already been said in Section 2.8 that rather than eliminating cautiously and finally, it is quicker and better to eliminate in a rather cavalier fashion in the first place and later to review the rejects in the improved light of the insight developed in the interim, a process here called repêchage because that is exactly what it is.

A very good test to use in the repêchage is the sort already mentioned in regard to particular options in Section 2.8, to set aside the weaknesses of the scheme in question and to decide whether there is any plea to be made on its behalf, any advantage it possesses, or any circumstances in which it would be viable.

Great importance should be given to these two ideas, of rough elimination and repêchage, and of weighing advantages in the first place regardless of disadvantages, which is why they have been repeated in such a short space. The former saves time and yet reduces the chance of overlooking a good solution: the latter inspires inventions. Simple and perfect ideas are as hard to find as metallic meteorites; it is better to take drossy ore and start smelting if you need iron.

Repêchage need not be confined to your own schemes; historical devices should also be re-examined. An example of this is the celebrated IBM 'golf-ball' typewriter. This incorporates all the type in hard plating on a plastic ball that is rotated in two angular degrees of freedom so as to bring any desired character opposite the paper. Now typewriters with all the type in a solid cylindrical piece, rotated and shifted along its axis to select the character, are, relatively speaking, of the highest antiquity, but have not survived for ordinary purposes. One of the chief weaknesses of the principle is the large inertia of the type, which has to be shifted bodily each time a character is typed; the ability to make it in a material of about a tenth the density has contributed largely to making this form of typewriter viable today. The IBM typewriter is full of excellent features, however, and the whole design is very refined. For example, of the two degrees of freedom the typing station must have relative to the paper, along the lines and from line to line, one is given to the paper and one to the type; this is fundamentally a better way, but not practical with the conventional typewriter, where both are given to the paper with rather clumsy results (check for yourself which has the fundamental advantage from the inertia point of view, the sphere or the cylinder).

More recently, another ancient form of typewriter has been revived, in which the type is arranged round the circumference of a disc which is deflected locally to force it against the paper when the required character has been rotated into the printing position. This 'daisy wheel' construction, so-called because the rim is divided into sectors like the disc of a daisy to make it easier to deflect locally, was used earlier this century only in very simple toy typewriters, but is now found in advanced machines.

It is not worth attempting to list exhaustively the changed aspects or circumstances which may result in reinstating a scheme which has been provisionally eliminated. Often the designer's view about which are the greatest difficulties of the problem will change, so that slight advantages become large ones, and large disadvantages become slight. Cheaper embodiments may be devised for what at first appeared too expensive a design, perhaps involving different materials or methods of manufacture. Most often, however, it is a new idea about how the scheme in question should be worked, sometimes a new idea which has arisen in connection with some other scheme and is then seen to have even more promise for one of the arrangements eliminated in the early rounds.

2.14   Combinative ideas: general remarks

In this chapter a sketch has been given of the scope and method of working of combinative approaches. Their virtues are that they provide a way of getting to grips with difficult problems, they offer some slight guarantee against overlooking very good solutions and a near certainty of finding good ones if they are there to find, and they evolve by stages more powerful special-purpose design tools suited to the problem in hand.

They proliferate embryo schemes, often in intractable numbers, which is a very good reason for keeping tables of options and so forth as small and as simple as possible without destroying their usefulness. Reduction of the number of 'schemes' to a manageable few can usually be partly done, and occasionally completely done, by systematic elimination, and parameters descriptive of the relative difficulty of functions can sometimes be of great help. Much of the time recourse must be had to *ad hoc* methods or engineering judgement, and at this stage synthetic approaches begin to be more profitable, though the combinative table can still be used with advantage as a source of material.

Together with the insight-developing kinds of study suggested in Chapter 4, combinative approaches provide something to work on, and with, and at. This work, combined with evolutionary devices such as hybridisation and redistribution of functions, can serve to built up synthesising principles and specific design philosophies, which are the last and most powerful aids to conceptual design.

Questions

Q.2.1. A range of high-tensile steel bolts are needed for use in bolting together steel structures such as frames of buildings, bridges, etc. These are to be friction-grip bolts, relying on clamping the surfaces of the members joined sufficiently hard together for the friction between them to carry the applied shear loads. Typically, the shank diameter is about 30 mm, and to guarantee sufficient residual load in the bolts it is necessary to take them into yield. Clearly, it is important to be able to ensure that the bolts are correctly tightened. Make a combinative study of this problem.

Q.2.2. Study the problem of powering deep-sea exploratory submarines.

Q.2.3. Consider the type of garden motor-mower which has a bladed cylinder rotating against a stationary blade and a driven roller at the rear—a typical small mower of the cylinder type, in fact. What arrangements of clutches are possible, and what are their merits?

Q.2.4. Make a classification of valve-operating gear for four-stroke petrol engines for private cars, from crankshaft to valve stem.

Q.2.5. The strain in the bore of aircraft gas turbine discs due to centrifugal loads is of the order of 0.003, so that a disc on a shaft of 100 mm diameter will be loose by 0.3 mm when running if nothing is done about it. List as many practicable methods of centring such discs as you can think of, and try to find a system of classification. Make your approach as abstract and exhaustive as you can.

Q.2.6. Analyse the problem of fore-and-aft stability of prams (baby carriages), and list the means of solution in as general and abstract a form as seems useful. Note any practical examples.

Q.2.7. Make a functional analysis and a table of options for kitchen tin-openers for opening cylindrical tins with the usual rolled rim joint.

Notes

For this sort of problem, complete answers are not possible. The following notes are offered.

N.2.1. The dominant functions are 'enabling tightening to the correct load' and 'ensuring tightening'. A torque spanner enables tightening to the correct load but does not ensure it; there is a premium on a bolt design which can be quickly inspected for correct tightening afterwards. It is valuable to consider as separate means of tightening imposing a given load (as with a torque spanner) or imposing a given distortion (as by turning $\frac{3}{4}$ turn after the nut tightens). Means of ensuring tightening may be by observing a given distortion. Then the distortion may be of the bolt shank itself, of the bolt-head, of the nut, of the washer or of some component used in tightening. In the Torshear bolt this is an extension of the bolt shank which is used to react the nut-tightening torque; tightening continues until this extension is twisted off, leaving clear evidence of correct tightening. In the GKN load-indicating bolt, the square head is undercut on its bearing face except for four triangular areas at the corners; under the correct tightening load the head partially collapses, reducing the gap between the undercut part of the underside of the bolt head and the parts joined. Gauging of this gap ensures the correct loading.

   An interesting recent design of load-indicating bolt is the RotaBolt, which has a hole extending down its axis from the head end[7]. A pin is screwed into this hole at the bottom, and threaded on the pin and trapped between its head and that of the

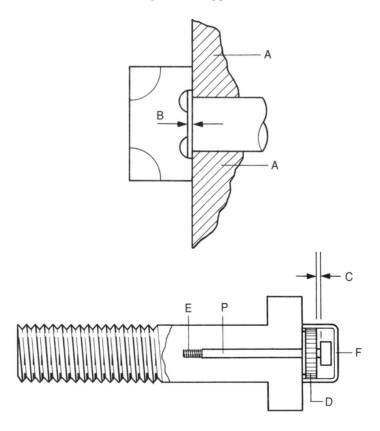

Figure 2.16   The square head of the GKN Load Indicating Bolt (*above*) tightens on four triangular pads A at the corners leaving a gap B elsewhere. Under the correct load, the head partially collapses, closing the gap B—which can be checked. The Rotabolt (*below*) has a central hole drilled through the head and into the shank, in which is put a headed pin P which is fixed at the inner end E. The disc D is free to turn on the pin because there is a small clearance at C. When the bolt is tightened and stretches, the gap C disappears and the disc D is no longer free to turn. A cap F protects the parts when not in use.

bolt is the load-indicator—a flat disc with a hole through it. Before the bolt is tightened, there is a little axial play for the disc between the two heads, and it turns freely on the pin. As the bolt is tightened its shank stretches, the play is taken up, and at the set load the disc is nipped between the two heads and will no longer turn freely. This design has the advantage that any relaxation of the load frees the indicator disc again, and so can easily be detected (Figure 2.16).

N.2.2. The parametric plot of Figure 2.17, which shows the viable regions for types of power supply for spacecraft, has some bearing on this problem.

N.2.3. Basically, there are three shafts—engine (E), roller (R) and cutter (C)—which can be connected with one or two clutches in the following ways, where X = clutch and ( ) denotes a solid shaft

One clutch   *Two clutches*

1 EX(CR) ⎫   both have merits, but the   4   EXCXR
2 (EC)XR ⎭   second has severe snags also   5   CXEXR

3 (ER)XC   this seems useless   6   EXRXC

Note that while 5 is more versatile thàn 4, it has the disadvantage that to declutch the engine entirely means operating two controls instead of merely one. The battle is between 1 and 4.

N.2.4. Some points of classification are:

   camshaft drive—bevels, helicals, spurs, chain, toothed belt, multiple
      eccentrics,
   rocker arms or direct operation,
   valve closure—by springs or 'positive' (desmodromic).

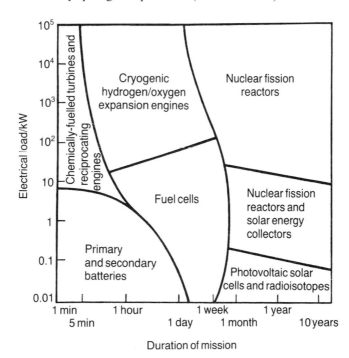

Figure 2.17   Plot of viable power sources for use in space. (By kind permission of Dr B.C. Lindley and the *Journal of the British Interplanetary Society*.)

N.2.5. See Section 7.4 and Figure 7.14.

N.2.6. High stability is safe, but makes it difficult to negotiate kerbs, because the pram is more difficult to tilt up to get the front wheels on the pavement. Radical solutions may be classified as:

(1) Low initial stability with high terminal stability, i.e. just prior to oversetting,

(2) Variable-geometry prams (the variation probably being achieved by the same operation as setting the brake, since brake-off/high-stability and brake-on/low-stability conditions are never needed).

Both solutions have been or are available.

N.2.7. For example,

Function	Options
Cutting	Wheel, knife
Driving cutter	Rollers, cutting wheel, toothed wheels, hand
Source of power	Motor, hand
Guiding cutter	Rim of tin, preset radius arm, hand
Retaining lid	None, magnet, sucker

(the first line contains an arbitrary decision).

3 Optimisation

3.1 Introduction

In the last chapter we considered the problem of how to select the best arrangement or combination of means for performing a given function. The variations under consideration were of kind, not of degree. But even when all decisions of kind have been made, so that all the qualitative aspects are settled, there remains the easier problem of fixing all the quantities involved. The solution of this problem we shall call 'optimisation'.

Before considering optimisation generally four examples will be examined to fix ideas and illustrate important points.

3.2 Making capital and running costs commensurate

In all our optimisations the ultimate object will be minimum cost, though sometimes a more immediate object like minimum weight may be substituted temporarily. In dealing with the structural aspects of aircraft engines, for example, it is almost possible to equate minimum cost with minimum weight, and the latter criterion is easier to work with.

Costs are of many kinds and given in different terms, and before their total can be minimised they must be put on a common footing. Associated with a prime mover, say, there may be a first cost, in pounds, a fuel cost, in pounds per tonne, a maintenance cost in pounds per 10 000 hours and so on, and the total cost may be expressed in pounds per kilowatt hour, or pounds per year. To combine the separate costs is generally simple enough but it depends upon other data, such as operating hours per year. In particular, capital cost depends upon many items, notably, interest and taxation rules, and is treated in practice in various ways, some of which are clearly inappropriate. For the present purpose, where it is necessary to make capital and running costs commensurate, it will be assumed that this is done by reducing capital cost to a fixed annual charge sufficient to pay back the capital over the life contemplated while yielding a stipulated percentage on the capital still outstanding. This is exactly what happens under the usual form of mortgage, where the householder pays an annual sum consisting of interest on the borrowed money outstanding and an element of repayment; the sum is the same every year, but is fixed at such a level that the money is all repaid by the end of a stipulated period. If

p is the interest, P the amount borrowed, and n the number of years for which the mortgage is to run, the fixed annual payment is kP, where

$$k = \frac{p}{1 - (1+p)^{-n}}$$

(check that this expression tends to the correct limit as $p \to 0$).

For example, if $n = 18$, $p = 0.10$ (i.e. 10 per cent), $k = 0.122$, so the householder pays £122 per annum for £1000 borrowed. The case of an industrial investment is similar, except that p is the return expected and n the expected life.

In many projects, where the return and costs are not fixed over the life, a more complicated treatment is needed, but it can be based on exactly the same idea (Discounted Cash Flow: see reference 42). Such complications and those introduced by taxation and capital gearing are beyond the scope of the present work.

3.3   Optimum speed of a tanker

We shall study in a very simplified form the question of the optimum speed for an oil tanker of given size, forming part of a fleet sufficiently large for the repercussions of this decision on it to be negligible.

We shall minimise the costs per tonne-kilometre of cargo carried. The *annual* costs can be divided into two categories, those which are independent of speed (total a) and those which are dependent on speed (total b). The most important items are:

a: cost of hull, fittings and some machinery, crew costs,
b: some of main engine costs, some machinery, fuel.

The drag of the ship is roughly proportional to v^2, where v is the speed, and so the power required is proportional to v^3. Thus the total annual costs can be written in the form

$$a + b \frac{v^3}{v_0^3} \tag{3.1}$$

where v_0 is some reference value of v for which the costs b are calculated. But the tonne-kilometres per year are proportional to v, so that the costs per tonne-kilometre are proportional to

$$\frac{annual\,costs}{v} = \frac{a}{v} + \frac{bv^2}{v_0^3} \tag{3.2}$$

The problem, then, is to choose v so as to make expression 3.2 a minimum, which is done by putting

$$\frac{\mathrm{d}}{\mathrm{d}v}\left(\frac{a}{v} + \frac{bv^2}{v_0^3}\right) = 0$$

or $\qquad -\dfrac{a}{v^2} + \dfrac{2bv}{v_0^3} = 0$

so that $\qquad v = \left(\dfrac{a}{2b}\right)^{1/3} v_0$ $\hspace{4cm}$ (3.3)

the required optimum or *cheapest* speed. Substituting this value of v in expression 3.2, it takes the form

$$\frac{a}{v_0}\left(\frac{2b}{a}\right)^{1/3} + \frac{b}{v_0}\left(\frac{a}{2b}\right)^{2/3} = (2+1)\frac{a^{2/3}b^{1/3}}{2^{2/3}v_0} \hspace{2cm} (3.4)$$

i.e. the speed-independent costs are twice as great as the speed-dependent ones at the optimum speed.

Suppose now a special tanker is built which is necessarily more expensive to build than the ordinary sort, but has ordinary propulsive machinery. If we held the speed at the customary value for the size of ship in question, the a-type, or speed-independent costs, would be greatly increased, leaving the b-type, or speed-dependent costs, unaltered, and the 2:1 ratio of the two types would be upset. From equation 3.3, we see that the speed should be increased as the third root of the a-type costs.

The liquid natural gas (LNG) tankers mentioned in the previous chapter exemplify this. Because of their expensive hull constructions and the high cost of the tanks, insulation and cargo-handling machinery, the a-type costs are increased by a factor of about 2 for a given hull size. If the speed were only the same as that of an ordinary tanker, costs per kilometre would go up from

$$\begin{aligned}
&2 \text{ units } (a\text{-type}) \quad + 1 \text{ unit } (b\text{-type}) \\
\text{to} \quad &4 \text{ units } (a\text{-type}) \quad + 1 \text{ unit } (b\text{-type})
\end{aligned}$$

i.e. from a total of 3 units to a total of 5 units.

However, from equation 3.3, we should put the speed up by a factor of $2^{1/3}$, and from expression 3.4 the costs become

$$2 \times 2^{2/3} \text{ units } (a\text{-type}) + 2^{2/3} \text{ units } (b\text{-type})$$

a total of $3 \times 2^{2/3} = 4.76$ units.

This illustrates a general principle of great importance, that any change in circumstances of a design ought to be followed through in all its consequences. The speed of an LNG tanker is about 17.2 knots, as against the 13.5 knots or so of a tanker of the same size, an increase of about 27 per cent ($2^{2/3} = 1.26$). However, there are other factors acting in favour of a higher speed for LNG tankers, besides high capital cost.

Figure 3.1 shows the variation of LNG transport costs with speed. Notice that in the above simple calculation, an increase in speed of 26 per cent reduced costs by less than 5 per cent (5 to 4.76 units) and the curves of Figure 3.1 show the same flatness.

It may be objected that the foregoing analysis is quite inadequate, that the representation of drag by a v^2 term is inaccurate, and so forth. This is so, but as a simple illustration it is valid enough. The two to one ratio at the optimum speed of the two large categories of cost may not be very precise, but it does show the kind of logic characteristic of situations of this sort.

Three points should be noted:

(1) Curves such as those in Figure 3.1 tell much more than just where the optimum is; they show how rapidly the cost rises as the optimum is departed from (see Section 3.10).

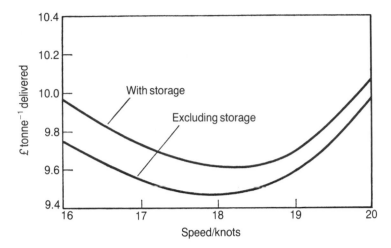

Figure 3.1   Speed of LNG tanker.

(2) The conditions determine the optimum; if the tanker problem is worked out with the tonne-kilometres per annum fixed instead of hull size, a slightly different balance is found, with a- and b-type costs in the ratio 7:3 instead of 2:1. Note that in this case M, the tonnage, is variable as well as v, the speed, but that Mv must be kept constant. The drag is proportional to area times v^2 or $M^{2/3}v^2$.

(3) Above all, the consequences of changes must be followed through in all their implications, if disadvantageous ones are to be mitigated and advantageous ones exploited to the full.

3.4   The optimisation of the sag:span ratio of a suspension bridge

Let us look very briefly at an example rather similar to the last before proceeding to a more difficult one. Figure 3.2 is a diagram of a suspension bridge of span s and central sag h. Suppose the deck to be of constant weight per unit length w, and to have side spans of length $s/2$, so that each tower carries a downward load ws (neglecting the weight of the cables and hangers). If $\frac{1}{2}T$ is the tension in each of the two cables, the vertical component of T must be $\frac{1}{2}ws$. Since the cable hangs in a parabola, the tangents PR, PS to it at the tower tops intersect at point P, h below the centre of the cable.

The vertical component of T is $T\sin\theta = \frac{1}{2}ws$. In practice, the sag is small, so

$$\sin\theta \simeq \frac{2h}{\frac{1}{2}s} = \frac{4h}{s}$$

and $$T \simeq \frac{ws^2}{8h}$$

The tension in the cables, and so their cross-sectional area, is thus inversely proportional to h. If we make the towers taller and h greater, the towers become dearer and the cables cheaper. Suppose f_c, c_c, f_t, c_t are the allowable stresses and *in situ* costs per unit volume of the cable and tower materials, respectively, we can write down a rough total cost of cables and towers like this. The combined cross-sectional area of the cables is

$$\frac{T}{f_c} \approx \frac{ws^2}{8hf_c}$$

their length is roughly $2s$, and their cost is approximately

$$\frac{ws^2}{8hf_c} \times 2s \times c_c = \frac{ws^3 c_c}{4hf_c}$$

The cross-sectional area of each tower is ws/f_t, its height is $(h + g)$ (Figure 3.2) and its cost is

$$\frac{ws}{f_t}(h+g)c_t$$

The total cost C_T of two towers and the cables is given by

$$C_T = 2ws(h+g)\frac{c_t}{f_t} + \frac{ws^3}{4h}\frac{c_c}{f_c} \tag{3.5}$$

To minimise C_T we put

$$\frac{dC_T}{dh} = 0 = 2ws\frac{c_t}{f_t} - \frac{ws^3}{4h^2}\frac{c_c}{f_c}$$

giving $\left(\dfrac{h}{s}\right)$ optimum $= \left(\dfrac{1}{8}\dfrac{c_c}{f_c}\dfrac{f_t}{c_t}\right)^{1/2}$ \hfill (3.6)

Putting some rough values in equation 3.6, say, $f_c = 600\,\text{N mm}^{-2}$, $f_t = 120\,\text{N mm}^{-2}$, and $c_c/c_t = 0.5$, gives $h/s = 1/9$, which corresponds roughly with present practice. Notice that c_t will be very high here (more than four times the raw-material cost) because of the costs of prefabrication, transport to the site and erection, the last two being a function of weight. Also c_c is only the marginal cost of adding a cubic metre more cable. Although high-yield steel will clearly pay its way for the towers, f_t for this purpose is the part of the allowable stress which can be allotted to taking direct load, since bending also occurs (see Section 6.10).

Substitution from equation 3.6 in equation 3.5 shows that in the optimised case, the cost of the cables is equal to the cost of the h or variable part of the towers (that above the lowest point of the cables; see Figure 3.2).

The simple treatment of equation 3.5 does not adequately represent the complexity of the problem. The weight of the cables contributes substantially to the

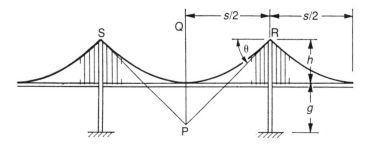

Figure 3.2   Suspension-bridge statics (vertical scale exaggerated).

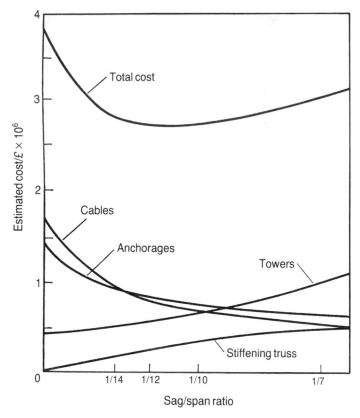

Figure 3.3   Suspension-bridge costs (original Severn design). (By kind permission of the Editor, *Structural Engineering*.)

load carried by the towers; increasing the height of the towers adds to their cost more than proportionately; the costs of the hangers by which the deck is suspended and the anchors at the ends of the cables need incorporation (the former will behave like tower cost, the latter, like cable cost); above all, the important problem of stiffness needs taking into account.

Stiffness in the vertical direction comes from two sources—the deck structure and the cables. The deck can be stiffened by attaching trusses, or it can form the top of a box girder. The cables are able to provide stiffness because of the heavy loads in them; to bend the deck it is necessary to alter the polygonal curve of the cables, and this requires forces which will be in direct proportion to the tensions in them. The lower the sag:span ratio, the higher the tensions in the cables and the less the need of a stiffening truss or stiff deck.

All this can be seen in Figure 3.3 where the problem we have treated very roughly has been done properly. It will be seen how the cable cost varies very nearly inversely as the sag, and the anchorages vary nearly in the same way but with a larger fixed term. The towers, however, increase in price somewhat more rapidly than proportionately to the sag.

3.5   Optimisation with more than one degree of freedom: heat exchanger

The two examples considered so far involve only one degree of freedom of choice—in the first case we could only vary the speed of the ship v, and in the second case we could only vary the sag:span ratio, h/s. In many practical cases we have more than one degree of freedom of choice, more than one thing we can vary to improve our criterion, here taken to be low cost.

Consider the case of a heat exchanger: we would like this to have zero capital cost, to produce no pressure drops, and to have a zero minimum temperature difference between the two sides. Compared with this imaginary device, a real exchanger involves three items of cost:

   (1) capital cost,
   (2) costs associated with the pressure drops,
   (3) costs associated with the temperature difference between the two sides.

To optimise the design we need to put all three kinds on a common footing, and how we do this will depend on the application.

Although a good example, this one tends to become thick with complicated algebra; moreover, not all readers will be familiar with heat transfer, so an unusual treatment will be adopted here which circumvents this difficulty.

To simplify the problem as far as possible, let us study only the tube side of a counter-flow heat exchanger in which a fluid of substantially constant properties is heated by a tube wall everywhere δT hotter than itself. The pressure drop in the fluid is Δp, and occurs entirely in the tubes. The total area of tube wall is A. The total cost C_T is put in the form

$$C_T = k_1\alpha + k_2\pi + k_3\tau \tag{3.7}$$

where k_1, k_2 and k_3 are constants,

$$\alpha = \frac{A}{A_0}, \pi = \frac{\Delta p}{p} \text{ and } \tau = \frac{\delta T}{T}$$

and A_0, p and T are a suitable area and the pressure and temperature at entry to the heat exchanger respectively. The three terms on the right-hand side of equation 3.7 are the capital cost, the cost of the pressure drop, and the cost of the temperature differential respectively, all reduced to some common basis such as annual cost, total cost over a given life, cost per kilogramme of fluid passing etc. (see Section 3.2).

In general we can reduce any of these terms at the expense of either of the others. By keeping A constant but making the tubes fewer in number and longer, we can increase the speed of flow and so decrease τ, but at the same time π will be increased. We can reduce both π and τ by increasing A in a suitable fashion, and so on.

Let α_0, π_0, τ_0 be the values of α, π, τ for some trial design. Consider the effect of increasing the number of tubes by a factor x and their length by a factor y. Then the area of tube-wall will be increased in the ratio xy, and α becomes $\alpha_0 xy$. The velocity of the fluid is reduced in the ratio $1/x$ since it is divided between x times as many tubes and the pressure drop per unit length of tube is proportional to the velocity to the power of 1.8 roughly, so that π becomes

$$\pi_0 x^{-1.8} y$$

Finally, the rate of heat transfer per unit area of tube wall is roughly proportional to the velocity to the power of 0.8, the area is up in the ratio xy and τ becomes

$$\frac{\tau_0 x^{0.8}}{xy} = \tau_0 x^{-0.2} y^{-1}$$

For a general design scaled from the trial one by the factors x, y,

$$C_T = k_1 \alpha_0 xy + k_2 \pi_0 x^{-1.8} y + k_3 \tau_0 x^{-0.2} y^{-1} \tag{3.8}$$

Now C_T is minimised with respect to x, y by putting

$$\frac{\partial C_T}{\partial x} = \frac{\partial C_T}{\partial y} = 0 \text{ (see Section 3.10) giving}$$

$$k_1 \alpha_0 y - 1.8 k_2 \pi_0 x^{-2.8} y - 0.2 k_3 \tau_0 x^{-1.2} y^{-1} = 0 \tag{3.9}$$

$$k_1 \alpha_0 x + k_2 \pi_0 x^{-1.8} - k_3 \tau_0 x^{-0.2} y^{-2} = 0 \tag{3.10}$$

Multiplying equation 3.9 by x and equation 3.10 by y and subtracting, an expression for y in terms of x is found:

$$y = \left(\frac{2}{7}\right)^{1/2} \left(\frac{k_3 \tau_0}{k_2 \pi_0}\right)^{1/2} x^{0.8} \tag{3.11}$$

Substituting for y in equation 3.9, we find

$$x = 2.5^{5/14} \left(\frac{k_2 \pi_0}{k_1 \alpha_0}\right)^{5/14} \tag{3.12}$$

Substituting for y and x in equation 3.8 we find for the minimum value of C_T

$$C_T(\min) = 2.7 \left(\frac{5}{14} + \frac{2}{14} + \frac{7}{14}\right) (k_1 \alpha_0)^{5/14} (k_2 \pi_0)^{2/14} (k_3 T_0)^{7/14} \tag{3.13}$$

where the fractions have been kept to show the division of C_T (min) into 5/14 attributable to capital cost, 1/7 attributable to pressure losses, and 1/2 due to temperature difference.

Figure 3.4 shows that large departures from the ideal distribution between capital, pressure drop and temperature difference costs are possible without much loss—for example, when all three are equal the total cost is about 10 per cent above the minimum.

There are many ways in which this analysis could be invalid in a given case: to mention a few, there may be a fundamental temperature mismatch (Section 5.7) the cost of end fittings has been ignored and the flow may tend to move out of the turbulent regime. The first two complicate the problem; the last can change its character, with the optimum at a discontinuity and not a turning value (Section 3.10).

The treatment of heat exchangers of all kinds, with changes of phase, variation of fluid properties, different configurations and so forth, will not be attempted here, but a few simple general observations are made in the next two sections.

It should be noted that the trick of using scale factors x and y separates the difficulties of the heat transfer aspect from those of the optimisation; the full value of this device is best appreciated by attempting the same problem without its aid.

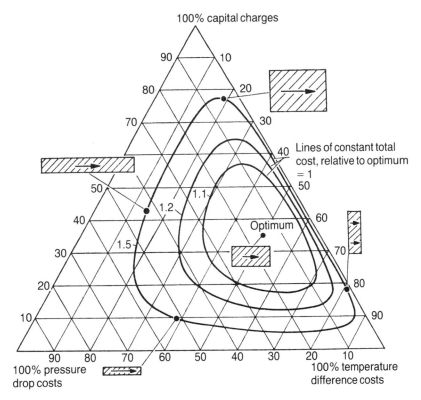

Figure 3.4   Total costs of heat exchanger. The hatched rectangles show the relative sizes of different designs.

3.6   Putting a price on the heat-exchanger performance

We have already noted the need to make capital and running costs commensurate. We now study the problem of putting a cost on performance figures such as π and τ, i.e. of giving values to k_1 and k_2 in equation 3.7. The procedure will be illustrated by reference to a simple open-cycle gas turbine, an example that brings out several important points.

Figure 3.5 is a flow diagram for a gas turbine in which the air leaving the compressor at temperature T_2 is heated in a counterflow heat exchanger by the exhaust gases. As the products of mass flow and specific heat for the air and the exhaust gases are nearly equal, an ideal heat exchanger would raise the air temperature to T_4, the temperature of the gases leaving the turbine. If T_3 is the turbine inlet temperature, the temperature rise through the combustion chamber would be $T_3 - T_4$. From all points of view, it is desirable to keep T_3 fixed at the highest value the blade material will permit. With a real heat exchanger the air will only be raised to $T_4 - \delta T$, where δT is the temperature difference in the heat exchanger. Since T_3 is fixed, the temperature rise through the combustion chamber will be $T_3 - T_4 + \delta T$. To achieve this, the fuel flow must be increased above that which would be necessary with an ideal exchanger, roughly in the ratio

$$\frac{T_3 - T_4 + \delta T}{T_3 - T_4},$$

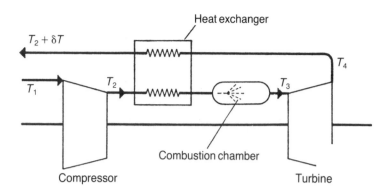

Figure 3.5   Gas-turbine flow diagram.

and it is a straightforward matter to put a cost on this additional fuel consumption.

It is not so simple to put a cost on Δp, the pressure drop through the heat-exchanger side. The effect of Δp is to reduce the pressure ratio R over the turbine, reducing the temperature drop $T_3 - T_4$ (since T_3 is held constant). The work done in the turbine per unit mass is reduced, and the output power is reduced by the same amount since the power required to drive the compressor is unaltered. Presuming the output power to be specified, we must increase the mass flow.

To estimate the decrease in $T_3 - T_4$, it is reasonable to assume that the polytropic efficiency of the turbine remains unchanged at η, say. Then

$$\frac{T_3}{T_4} = R^{\eta \frac{\gamma-1}{\gamma}}$$

where γ is the ratio of specific heats. This equation may be written

$$\ln T_3 - \ln T_4 = \eta \, \frac{\gamma - 1}{\gamma} \ln R$$

Differentiating with respect to R, since T_3 is constant,

$$-\frac{\mathrm{d}T_4}{T_4} = \eta \frac{\gamma - 1}{\gamma} \frac{\mathrm{d}R}{R}$$

and $\dfrac{\mathrm{d}R}{R} \simeq -\dfrac{\Delta p}{p}$

where p is the mean pressure in the side of the heat exchanger being considered.

The $\Delta p / p$ increases by 1 per cent in a case in which $T_4 = 0.7\, T_3$, $\eta = 0.84$ and $\gamma = 1.4$,

$$\frac{\mathrm{d}T_4}{T_4} = \frac{0.84(1.4 - 1)}{1.4} \times 1\% = 0.24\%$$

and the temperature drop through the turbine decreases by a fraction

$$\frac{T_4}{T_3 - T_4} \times \frac{\mathrm{d}T_4}{T_4} = \frac{0.7}{0.3} \times 0.24\% = 0.56\%$$

This is the drop in turbine shaft power. The output power may be 45 per cent of the turbine power, perhaps, the remainder driving the compressor. In such a case the drop in output power is

$$\frac{100}{45} \times 0.56\% = 1.24\%$$

To restore this output power drop requires the same increase of mass flow.

The rise in fuel consumption will not be quite as much as 1.24 per cent because the rise in T_4 produces an increase in the temperature of the air leaving the high-pressure side of the heat exchanger; as a result the temperature rise in the combustion chamber is less. In a typical case, the increase in fuel consumption might be about 0.9 per cent.

Having found that, if the power output is to be maintained, an additional pressure drop of 1 per cent through the high-pressure side of the heat exchanger means an increase of about 0.9 per cent in fuel consumption and about 1.2 per cent in mass flow (and hence of machine size generally), we can put a cost on these penalties, say, in pence per kilowatt hour. Naturally, the pressure drop and temperature difference in the other side of the heat exchanger must be included in a full treatment.

The approach we have used to find the effects of a small increase of pressure drop is of a kind called here a 'differential study'. It is quick, accurate and cheap, and excellent for developing insight (Section 4.3).

3.7   Variation of costs with application

As with the tanker speed problem, it is important to recognise that a change in any k in equation 3.7 will not affect the relative sizes of the terms $k_1\alpha$, $k_2\pi$ and $k_3\tau$ in the optimised case, but will affect the relative sizes of α, π and τ. For example, in a prime mover, τ often involves only higher fuel consumption, whereas π involves both higher fuel consumption and higher first cost. In a mechanical refrigerator, where both π and τ demand increased inputs of mechanical work, they are on a more equal footing and will tend to appear in the proportions that fundamental thermodynamics suggests. For a refrigerator, we can almost write, instead of equation 3.7,

$$C_T = k_1\alpha + k_4\Delta s$$

where Δs is the increase in specific entropy through the exchanger (since there is no net change of enthalpy in the exchanger, the increase in entropy in it multiplied by the ambient temperature is the lost work). We must expect the ratio of π to τ to be higher in the exchangers of an air separation plant, say, than in those of a gas turbine.

3.8   Further aspects of heat-exchanger optimisation

Before leaving the heat-exchanger problem, there are two important points to be mentioned briefly.

The possibility of varying element size has been considered, but usually this aspect can be settled separately. Sometimes the answer is, the smaller the diameter of the tubes the better, within the limits set by fouling. Sometimes the cost of end joints is dominant, or perhaps limits are imposed on the fluid velocity.

The question of element type is also important. In some cases plain tubes, or perhaps tubes with integral fins, are the only acceptable solution, limitations being imposed by pressure, thermal shock etc. In other cases a wide choice of elements are available. On the whole, the more elaborate forms of element increase both the local heat-transfer rate and the resistance to flow, and the balance of these conflicting effects may swing either way according to the relative sizes of the k's and the properties of the fluids and the solid involved [11].

3.9   An elementary programming problem

Consider four parts W, X, Y, Z, which are used together in two alternative combinations,

one X, two Ys and one Z: arrangement A
or one X, one Y and one W: arrangement B

In each combination there is one critical fit involving several toleranced dimensions, one from each part; we shall use the symbols W, X, Y, Z to denote the tolerances on these dimensions, and W_0, X_0, Y_0, Z_0 to denote large reference values of the tolerances. We shall also write

$$W = W_0 - w, \; X = X_0 - x, \text{etc.}$$

so that w, for example, represents the amount we have tightened the tolerance from the rather open reference value W_0. For each complete set of parts to be produced, suppose the cost of this reduction w in tolerance is $2w$ pence, i.e. w costs 2p a unit. Suppose also that x costs 6p a unit, y costs 9p a unit and z costs 2p a unit. Then the total cost of tightening all the tolerances by w, x, y, z, respectively, is

$$C_T = 2w + 6x + 9y + 2z \, \text{p}$$

To fix ideas, the units might be thousandths of inches. If the tolerances W_0, X_0, Y_0, Z_0 are 1.5, 1.0, 1.0 and 1.5 units, respectively, then the range of fit that they would produce with arrangement A as between maximum and minimum metal conditions would be

$$X_0 + 2Y_0 + Z_0 = 4.5$$

and with arrangement B

$$X_0 + Y_0 + W_0 = 3.5$$

Suppose now that for functional reasons these ranges of fit are too great, and we need to reduce them to 2.5 and 2.0, respectively, so that

$$X + 2Y + Z = X_0 - x + 2(Y_0 - y) + Z_0 - z$$
$$= (4.5 - x - 2y - z) = 2.5$$

or $x + 2y + z = 2.0$

and similarly, $X + Y + W = 2.0$, reducing to $x + y + w = 1.5$

We require to satisfy these two equations with values of w, x, y, z that make C_T as small as possible; to make a reasonable problem, however, we must limit the values of w, x, y, z. Clearly, $W = W_0 - w$ cannot be indefinitely small; let us limit it to 0.8, so that $w \leqslant 0.7$. Let us impose also the limits $X \geqslant 0.3$, $Y \geqslant 0.5$, $Z \geqslant 0.5$, and write down the entire problem in formal mathematics:

Minimise

$$C_T = 2w + 6x + 9y + 2z$$

subject to the equations

$$\left. \begin{array}{l} x + 2y + z = 2.0 \\ x + y + w = 1.5 \end{array} \right\} \tag{3.14}$$

and the inequalities

$$w \leqslant 0.7, x \leqslant 0.7, y \leqslant 0.5, z \leqslant 1.0$$

This is a very simple problem in *linear programming* (see next section).

By means of the equations, we eliminate w and z from the expression for C_T and the inequalities, giving

$$C_T = 7.0 + 2x + 3y \tag{3.15}$$

$$\left. \begin{array}{l} w = 1.5 - x - y \leqslant 0.7, x + y \geqslant 0.8 \\ z = 2.0 - x - 2y \leqslant 1.0, x + 2y \geqslant 1.0 \end{array} \right\} \tag{3.16}$$

There are now four inequalities governing x and y, two in 3.14 and two in 3.16.

On an x, y plot each of these inequalities can be represented by a straight line (see Figure 3.6) and all four are satisfied by any point in the interior of the quadrilateral so formed, and only by such points. The region in which all the imposed conditions or 'constraints' are satisfied is called the feasible area, and in this case it is a quadrilateral (ABCD in Figure 3.6). (Notice that if the straight line AD were much lower in Figure 3.6, it would not be an effective constraint and the feasible area would be a triangle.) The problem reduces to finding the point in the feasible area for which C_T is least.

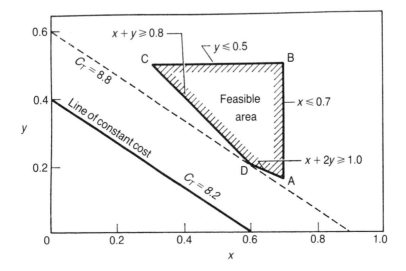

Figure 3.6   Tolerancing problem.

A constant value of C_T corresponds to a line of slope $-2/3$ (see equation 3.15). Thus we have only to draw the line of this slope and nearest to the origin which cuts the feasible area. Thus D ($x = 0.6$, $y = 0.2$) is the required solution, giving $w = 0.7$, $z = 1.0$ and $C_T = 8.8$.

Serious linear programming problems are much more elaborate, and require special methods for their solution.

There is a useful approach to a very easy problem like this one, the task/cost viewpoint. The task is to effect a total reduction of range of fit of

$$(4.5 + 3.5) - (2.5 + 2.0) = 3.5$$

Now a unit of x costs 6p, and reduces both ranges of fit (i.e. in arrangements A and B), a task/cost ratio of $\frac{2}{6} = \frac{1}{3} 10^{-3}$ in/p. A unit of y costs 9p and reduces one range of fit by 2 and the other by 1, giving the same task/cost ratio, and so on. We have

task/cost ratio, 10^{-3} in/p

w	$\frac{1}{2}$
x	$\frac{1}{3}$
y	$\frac{1}{3}$
z	$\frac{1}{2}$

Clearly, w and z are the best value for money and we 'buy' all we can of these, 0.7 of w and 1.0 of z for 3.4p. The remaining $(3.5 - 1.7)$ or 1.8 units we have to take in x and y, at a cost of $1.8/\frac{1}{3} = 5.4$p, giving a minimum C_T of 8.8p.

3.10   Classification of optimisation problems and methods of solution

Having studied four cases of optimisation, the simple ones of the speed of the tanker and the sag:span ratio of the suspension bridge, the two degree of freedom example of the heat exchanger and a simple programming problem, let us glance briefly at the whole field.

Problems can be classified in two ways:

(a) by the number of degrees of freedom, and
(b) by the kind of maximum.

Figure 3.7a shows three functions of x, y_1, y_2 and y_3 : y_1 has a maximum of the classical sort at M, found by putting $dy_1/dx = 0$; y_2 has a maximum at N, but dy_2/dx is not zero there; finally, y_3 has a maximum of the classical kind at P, but the maximum value in the range shown occurs at Q.

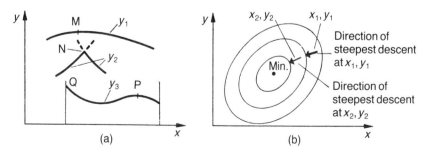

Figure 3.7   (a) Kinds of maximum, (b) Method of steepest descents.

Problems where a maximum such as M or P is sought are called turning value problems; problems where a maximum such as N or Q is sought are called programming problems. In practical situations it is not always possible to tell beforehand which kind of maximum will prove to be concerned (see Question 3.5).

In principle, programming problems are much the easier kind to solve, but in practice they generally have such large numbers of degrees of freedom as to make them very difficult. Their solution is beyond our scope here, but there is an abundance of literature [12, 13]. A few remarks on the solution of turning value problems will be made.

(a) Turning value problems with one degree of freedom

If $C_T(x)$ is a total cost, the minimum is found by putting $dC_T/dx = 0$. It is seldom if ever necessary to use the analytical tests, $d^2C_T/dx^2 > 0$ for a minimum, $d^2C_T/dx^2 < 0$ for a maximum because the sense of the problem will serve instead. In the case of the tanker,

$$C_T = \frac{a}{v} + bv^2$$

tends to infinity both as v approaches 0 and as v approaches infinity, for very real reasons, so that the intervening turning value must be a minimum.

If C_T cannot usefully be differentiated or the resulting equation readily solved, it may be best to evaluate C_T at a number of points. Special rules have been found for doing this economically (e.g. Fibonacci search), but these are rarely of use. In the problem of Section 6.4, the disc-brake caliper, it is more useful to plot the function concerned that just to find its minimum, and this is often the case.

When the value of the independent variable is restricted to certain fixed values, integers perhaps, the turning value is found as though this restriction did not exist and then the nearest permitted values of the independent variable on either side examined to see which gives the practical minimum. Thus, in Question 3.8 the independent variable is the number of cylinders in a pump, which is restricted to integral values.

(b) Turning value problems with more than one degree of freedom

If $C_T(x,y)$ is a total cost, we find the values of x and y which make C_T a minimum by putting

$$\frac{\partial C_T}{\partial x} = 0, \quad \frac{\partial C_T}{\partial y} = 0,$$

and testing the turning values so found, if necessary, by the tests

(1) $$\left(\frac{\partial^2 C_T}{\partial x \partial y}\right)^2 < \frac{\partial^2 C_T}{\partial x^2} \cdot \frac{\partial^2 C_T}{\partial y^2}$$

for a minimum or maximum, as against a saddle

(2) $$\frac{\partial^2 C_T}{\partial x^2} > 0, \quad \frac{\partial^2 C_T}{\partial y^2} > 0 \text{ for a minimum}$$

(3) $\dfrac{\partial^2 C_T}{\partial x^2} < 0,\ \dfrac{\partial^2 C_T}{\partial y^2} < 0$ for a maximum

(either half of conditions (2) and (3) implies the other half if (1) is established).

Only the simplest problems can be treated in this way, and more generally useful techniques are a specialised topic which cannot be dealt with here. The most celebrated is called 'hill-climbing' or the 'method of steepest descent'. A man trying to climb a hill in a fog, if he went always up the slope the steepest way, would be bound to arrive at the top—some sort of top, anyway. If C_T is the altitude above the x, y plane, then

$$\frac{dy}{dx} = \frac{\partial C_T / \partial y}{\partial C_T / \partial x}$$

is the compass bearing of the direction of steepest ascent (Figure 3.7b).

A point x_1, y_1 is chosen and the direction of steepest ascent found. To find the minimum, a further point x_2, y_2 is taken a suitable distance from x_1, y_1 in the reverse direction, that of steepest descent, and the process is repeated. It sounds easy enough, but in practice there are often grave difficulties and many variations on the basic procedure have been devised.

With more than two independent variables the principles of finding minima and maxima are not greatly altered but the practical difficulties increase rapidly.

When there are only two independent variables and the problem is an ordinary well-behaved engineering one, a brutal process of calculating C_T for various sets of values of x and y is often justified. This is what has been done in Figures 3.1 and 3.8. With a computer the cost may not be too high and the information contained in such plots is a great deal more useful than a mere knowledge of the minimum.

(c) With several variables and constraining equations

Consider the problem of finding the minimum of

$$C_T = k_1 M + k_4 M^{2/3} v^3 \tag{3.17}$$

where M and v are variable but subject to the constraining equation

$$Mv - Q = 0 \tag{3.18}$$

We could eliminate either M or v from equation 3.17 by means of equation 3.18, and then proceed in the usual way for a case of one independent variable. Another way is to use a Lagrangian or undetermined multiplier, λ, in the following fashion. We write

$$C_T = k_1 M + k_4 M^{2/3} v^3 + \lambda(Mv - Q) \tag{3.19}$$

which is obtained by adding λ times equation 3.18 to equation 3.17, and then proceed as for two *independent* variables M, v, treating λ as a constant. We find that

$$\frac{\partial C_T}{\partial M} = k_1 + \frac{2}{3} k_4 M^{-1/3} v^3 + \lambda v = 0 \tag{3.20}$$

$$\frac{\partial C_T}{\partial v} = 3 k_4 M^{2/3} v^2 + \lambda M = 0 \tag{3.21}$$

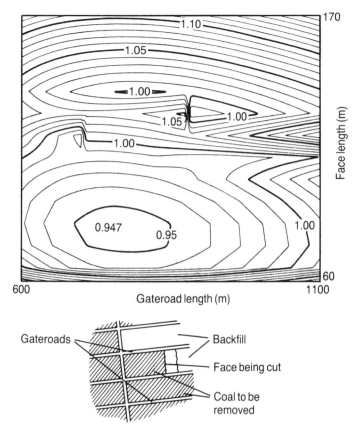

Figure 3.8 Optimisation of parameters in coal mining. Here the cost per ton of coal won is plotted against the length of face being cut and the length of the gate roads (see small figure) in retreat mining with reusable props. (By kind permission of the National Coal Board.)

Eliminating λ by subtracting v/M times equation 3.21 from equation 3.20 gives

$$k_1 = \frac{7}{3} k_4 M^{-1/3} v^3$$

which together with equation 3.18 yields the required minimum.

This device can be used with larger numbers of variables and equations. For example, to minimise $C_T(u, v, w, x)$ subject to the constraints $f(u, v, w, x) = 0$ and $g(u, v, w, x) = 0$, we can put

$$C_T = C_T - \lambda f - \mu g$$

differentiate partially with respect to u, v, w, x in turn and put each of the four partial derivatives equal to zero. Eliminating λ and μ gives us two equations in u, v, w, x which together with $f = 0$ and $g = 0$ enable us to locate the turning values of C_T.

A further example is given in Section 4.3. No proof will be given of this elegant device, which is sometimes useful.

(d) With an infinite number of degrees of freedom

A problem of historical importance was this; given two points, A and B, B lower than A, join them with a smooth wire such that a small bead, threaded on it and released from A, reaches B in the shortest possible time. This is a minimisation problem in which we have to find, not the values of a finite number of variables u, v, w etc. but a shape or function, the shape of the wire. This *brachistochrone* (Greek short time) problem is an optimisation problem with an infinite number of degrees of freedom.

The appropriate analytical tool for this problem is the 'calculus of variations' (see for example reference 14), which gives a differential equation for the form of the wire. This kind of optimisation rarely arises in engineering problems.

Sometimes a physical idea will lead to an alternative means of solving such problems, as in the next section and the case mentioned in Section 4.4.

3.11   The design of rotating discs: an optimum structure

(1) The non-rotating disc

Suppose we need a structure to resist a uniform radial outward load $2\pi P_0$ around the periphery of a circle of radius a, as in Figure 3.9. Stresses are to be based on a maximum shear stress criterion (Tresca) with an allowable tensile stress f. The allowable combinations of plane principal stresses are shown in Figure 3.10.

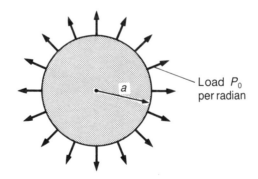

Load P_0
per radian

Figure 3.9   Uniformly distributed radial loading.

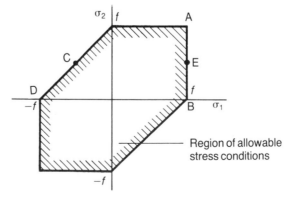

Region of allowable
stress conditions

Figure 3.10   Allowable stress, maximum shear stress (Tresca) criterion.

Such a load can be sustained by a disc of uniform thickness t, which will be stretched like a drumskin equally in all directions. By equating the load to the inward force at the rim and putting the radial stress equal to f,

$$2\pi P_0 = 2\pi a t f$$

and $t = \dfrac{P_0}{af}$

There is a *load diffusion problem* in getting the load into the disc, requiring a rim, but we shall ignore this.

This simple disc is, in fact, a structure of minimum weight. The volume of material in it is

$$\frac{\pi P_0 a}{f}$$

The principal stresses at any point in it are $(f, f, 0)$ so the material is all working at point A (Figure 3.10).

Now consider a set of spokes joining at the centre (let us ignore how this is to be done). Their combined cross-sectional area carries the load $2\pi P_0$ and so is $2\pi P_0/f$. The volume of material in the spokes is

$$\frac{2\pi P_0 a}{f}$$

i.e. just twice that in the disc. All the material has principal stresses in it of $(f, 0, 0)$, i.e. it is working at point B of Figure 3.10. The material of the disc is doing 'holding together' per unit volume of $2f$ against f for the spokes. The 'structural task' to be performed (Section 8.1) is $2\pi P_0 a$, and dividing it by the appropriate 'holding together per unit volume' gives the respective volumes of material required.

A third possible structure is a ring of radius a. By considering the equilibrium of a small arc (Figure 3.11) it is found that the force F in the ring is P_0 and hence the section area must be P_0/f and the volume of material

$$\frac{2\pi P_0 a}{f}$$

a result we could have predicted from the fact that the material is stressed in one direction only.

Of all the permissible working points in Figure 3.10, A has the maximum

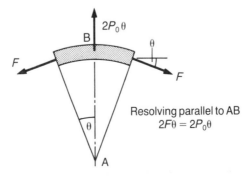

Figure 3.11 Element of ring.

'holding together' per unit volume, and all the material in the disc is working at it: the structure is therefore one of minimum weight.

(2) The effect of centrifugal forces

The ideal form of structure to sustain the load system of Figure 3.9 when there are no loads due to the material itself has been shown to be the flat disc of thickness

$$t = \frac{P_0}{af}$$

The commonest occurrence of large-load systems of this sort, however, is in rotating discs, and there the centrifugal forces on the disc material make the problem more complicated. It can however be solved very readily by a small extension of the original treatment.

If the angular velocity of the disc is ω and the density of the material is ρ, then the centrifugal load per unit volume on material at radius r is $\rho\omega^2 r$. The volume of disc lying between radii r and $r + dr$ is $2\pi rtdr$ (where t, the disc thickness is no longer constant but an unknown function of r).

The total centrifugal load on this elementary ring of material is $2\pi\rho\omega^2 r^2 tdr$. Now this load is uniformly distributed round a circle of radius r, so that it is the same kind of load system as that of Figure 3.9 and we have already seen that the ideal structure to carry such a load system is a disc of thickness

$$\frac{\rho\omega^2 r^2 tdr}{rf} = \frac{\rho\omega^2 rtdr}{f}$$

Thus at the inside edge of the elementary ring the disc needs to be just this amount thicker than at the outer edge. In other words, the *decrease* in thickness from r to $r + dr$ is

$$-dt = \frac{\rho\omega^2 rtdr}{f} \tag{3.22}$$

and this same relation must hold over any other elementary ring. Writing equation 3.22 in the form

$$-\frac{1}{t}\frac{dt}{dr} = \frac{\rho\omega^2 r}{f}$$

and integrating

$$-\ln t = \frac{\rho\omega^2 r^2}{2f} + \text{a constant.} \tag{3.23}$$

Putting in the end condition $t = P_0/af$ at $r = a$,

$$t = \frac{P_0}{af}\exp\left\{\frac{\rho\omega^2}{2f}(a^2 - r^2)\right\} \tag{3.24}$$

giving the kind of profile shown in Figure 3.12. This design also is of minimum weight.

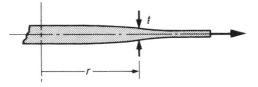

Figure 3.12   Uniformly stressed disc.

Optimising principles

The formal statement of the problem we have solved is: given the loading system of Figure 3.9, rotating with angular velocity ω, find the structure of minimum weight that will support it, subject to the allowable stress condition of Figure 3.10. This is a problem of determining an optimum shape, what we called in Section 3.10 a brachistochrone type of problem, and we have solved it without using any advanced mathematics. We have been able to do this because of a special property of the problem; there exists a measure $(2\pi P_0 a)$ of the structural task to be performed. If σ_1, σ_2 are the principal stresses in the disc (regarded as thin, so that the axial stress is zero) then in the case of the non-rotating disc

$$\int (\sigma_1 + \sigma_2)dv = 2\pi P_0 a$$

where dv is an element of volume and the integral is taken over the whole volume of the structure. Thus we have only to make $(\sigma_1 + \sigma_2)$ take everywhere the maximum permissible value by making the whole of the material work at point A of Figure 3.10 and we shall have a minimum-volume design. This sort of rule for obtaining an optimum design we shall call an autonomous local optimising principle, or ALOP for short—autonomous and local because it can be applied at a point regardless of anything that happens elsewhere. Such principles are of little practical use, but may be powerful aids to developing insight.

It is not always possible to satisfy the ALOP at all points, as in the problem of the next section.

3.12   Hub design

To complete this brief study of an optimal structure, the rotating disc, we shall consider the common case where a central hole of radius b must be provided. For simplicity the centrifugal forces will be regarded as negligible so that the ideal structure (in the absence of a central hole) is the uniform disc of thickness t.

Elastic design

By our earlier arguments we want to keep each point of the original disc in the same stress condition $(f, f, 0)$, and this means it will have to exert a radial force tf per unit length or bft per radian round the edge of the hole, as in Figure 3.13. Some structure will have to be provided to carry this load: in the unbored disc it was the central portion and this is the ideal structure for the purpose. Of the alternative structures we have studied only the ring is suitable, so this is the first thing to try. (The disc will be assumed to be thin, so that the radial extent of a suitable ring is small.) If the ring

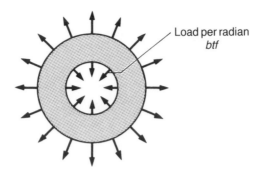

Figure 3.13   Loading of hub of disc.

is of cross-section A_h, then the hoop stress σ_h in it is

$$\sigma_h = \frac{\text{load per radian}}{\text{area}} = \frac{btf}{A_h} \text{(see Section 3.11)}$$ (3.25)

It will not do, however, to put $\sigma_h = f$ if we want the stress to remain everywhere less than f, i.e. if the disc is to be *elastically* designed. There is a relationship between the circumferential strain in the ring (henceforth called the hub) and the stress in it, i.e.

$$\text{circumferential strain in hub} = \frac{\sigma_h}{E}$$ (3.26)

When the desired stress condition $(f, f, 0)$ holds in the disc (henceforth called the web), the circumferential strain in it is everywhere

$$\frac{f}{E} - \frac{vf}{E} = (1-v)\frac{f}{E}$$

where v is Poisson's ratio. For the hub to match this the two circumferential strains must be the same, i.e.

$$\frac{\sigma_h}{E} = (1-v)\frac{f}{E}$$ (3.27)

and so from equation 3.25

$$A_h = \frac{bt}{1-v}$$ (3.28)

If A_h has some other value than this, the circumferential strain in the hub will match that in the web—it must as they are adjacent portions of the same piece of metal—but the stress condition in the web will not be the desired $(f, f, 0)$. Figure 3.14 shows the stress distribution in a disc in which the value of A_h is less than that for the elastic design (curves II) and it will be seen that the hoop stress exceeds f throughout. Figure 3.14 also shows the stresses in an *elastically* designed disc (i.e. with hub cross-section given by equation 3.28). The stress condition in the hub of this disc ($\{1-v\}f, 0, 0$) seems uneconomically low, and it may be suggested that if the reinforcing (hub) material were extended to larger radii it might be more highly stressed and result in a lighter design, still without exceeding the allowable stress f. It is not so.

   The elastically designed disc, with hub cross-section given by $A_h = bt/1-v$, is a

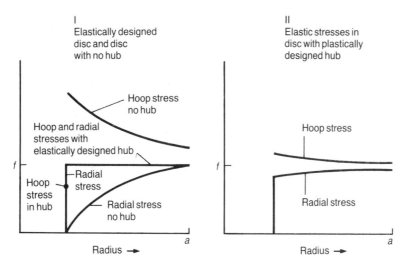

Figure 3.14   Elastic stresses in rotating discs.

minimum weight structure design to the allowable elastic stress condition of Figure 3.10. In a *plastically designed* disc, we keep the *plastic* stresses within the same boundary, but the strain is assumed to be plastic not elastic, so that the elastic compatibility equation 3.27 need not be met. We put $\sigma_h = f$ in equation 3.25 (which is one of equilibrium), so that $A_h = bt$, giving the elastic stresses we have already seen in Figure 3.14 (curves II).

Plastic design

Suppose the peripheral load on the disc is due to centrifugal forces on blading. Imagine the speed increased from ω to $k^{1/2}\omega$, so that all the loads, being proportional to the square of the speed, are increased by a factor k where

$$k = \frac{\text{u.t.s. of material}}{f}$$

Then $k^{1/2}\omega$ is about the speed at which the elastically designed disc will burst and k is an ultimate factor at speed ω.

Now when overspeeded a disc may burst in either of the patterns shown in Figure 3.15(a) and (b), and the circumferential failure in pattern (a) may have any radius r

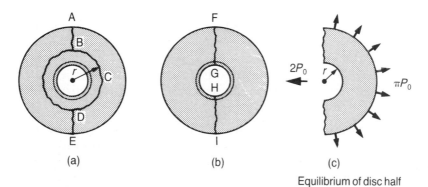

Figure 3.15   Plastic failure of discs.

within the web. The idea of plastic design is to make all these failures equally likely, an old principle exemplified by the one-horse shay (which, according to tradition, fell to pieces on its hundredth birthday). We assume that the material is ductile enough to ensure that, before failure takes place, the stress on the failing section will become practically uniform by plastic flow. The design principle then reduces to making the average normal stress on any section such as ABCDE or FGHI the same. In this case the average normal stress must be f at normal speed if we require an ultimate factor of k (average here means average with respect to the projected area on the diametral plane AE or FI; this assumes that there are no shear stresses on the section, which follows from symmetry). This is a simple equilibrium condition, and in the case where centrifugal forces on the material of the disc are negligible it reduces to (Figure 3.15(c))

$$2P_0 = \text{projected area of section} \times f,$$

i.e. the projected area of section is constant at $2P_0/f$. Now the projected area of any section such as ABCDE is constant at $2af$ if t, the thickness, is constant, so the web is uniform and

$$t = \frac{P_0}{fa}$$

The projected area of section FG–HI is

$$2t(a - b) + 2A_h = \frac{2P_0}{f}$$

so that $A_h = bt$ and the plastic design is complete.

If the centrifugal forces on the material must be taken into account, the thickness of the web can be shown to be given by equation 3.24, just as in the elastic design, and

$$A_h = bt \frac{f}{f - \rho\omega^2 b^2} \tag{3.29}$$

The term $f - \rho\omega^2 b^2$ in equation 3.29 represents a valuable concept we shall not discuss at length. It is the 'effective hoop strength' obtained by subtracting from f the stress needed to support the centrifugal forces on the material itself. An alternative method of deriving equation 3.24 can be based on this measure of usefulness.

The methods described enable the plastic and elastic design of discs to be done very simply and with adequate accuracy provided the thickness is small compared with the diameter. Slight modifications are needed in practice to deal with the rim, which is usually thicker than the web, to enable the load diffusion from the blades to be achieved with joints of practicable efficiency, and the finite radial extent of the hub. The design of discs is more difficult if the material properties vary with the radius or large thermal stresses are present, or if the form is relatively thick.

Points to note are:

(1) The forward process of designing a disc to have a given stress distribution in it is much less difficult than the reverse analytical process of finding the stresses in an arbitrary design, so long as the designed pattern of stress has a particular form which is also the most desirable in the majority of cases. It is the purposeful nature of the disc which makes it easier to design (see also Section

4.4). However, the last step in the design involves approximating to the ideal shape with one cheap to manufacture (the form of equation 3.24 hardly qualifies) and if necessary this shape must be stressed by an analytical method. (2) The design proceeds by the *matching* together of three components, the rim (which we have omitted), the web (with which we have dealt fully), and the hub (with which we have dealt rather inadequately). In this respect the problem of disc design is rather like one of systems engineering.

3.13   Summary

The most useful single approach to optimisation is probably the study of the effects of varying a few major parameters, as in Figure 3.3, without any calculus at all. Fortunately this unsophisticated but effective method is practical with the low cost of number-crunching on a computer. Even so, it is possible to struggle in confusion for a long time if the number of parameters is more than two, and then the use of more refined methods of the kinds outlined may bring enlightenment. Above all, where a simple analytical formulation of the cost is possible, even if it is rather rough, it may produce important insights, and save much work or lead to radical improvements.

Questions

Q.3.1(2). Using the same assumptions as Section 3.3 (e.g. drag proportional to v^2) prove the result given there for a ship optimised for a fixed number of tonne-kilometres per year, i.e. the ratio of a to b-type costs is 7:3.

Q.3.2(3). For a ship of given tonnage, on the assumptions of Section 3.3, what is the effect on optimum speed of allowing for a turn-round time t per voyage of length s? Will it be higher or lower than for the case $t = 0$?

Q.3.3(4). For a given number of tonne-kilometres per year, will the optimum speed of a ship be higher or lower if a turn-round time t per voyage of length s is allowed for, compared with the case $t = 0$? Answer for the two cases (a) t constant and (b) t proportionate to tonnage M, and for both state the nature of the critical term in the costs which is responsible for the result in question. (This question and Q.3.2 can be answered by reasoning alone, but a very high standard of insight is required, or else hindsight after doing the calculations.)

Q.3.4(2). Electrical power P is to be supplied at a DC voltage of about V over a distance s, a return cable being used. The conductors are of specific resistivity ρ and cost c_1 per unit volume year. Power costs are c_2 per unit year. Find the minimum total cost, assuming this to be small compared with that of the power transmitted.

Q.3.5(3). A uniform beam AB, pinned at each end and of length L, carries a uniformly distributed load w per unit length and a concentrated load wLy at a point C such that $AC = Lx$. Plot on an x, y graph the zone in which the maximum bending moment occurs at C.

Q.3.6(2). Assuming the validity of equation 3.8, what percentage increase in capital cost of a heat exchanger is necessary to reduce the temperature difference

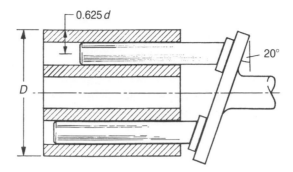

Figure 3.16   Swashplate pump.

by 1 per cent, without changing the pressure drop?

Q.3.7(4). Two depots B and C are each to be supplied with oil at the same steady rate from a third depot A; A, B and C form an equilateral triangle. A pipeline is to be laid from A to some point D on the perpendicular bisector of BC. From D two smaller pipes are to be led to B and C. Find the value of the angle BDC which minimises the sum of pumping and pipe costs, on the following assumptions:

Pumping costs are proportional to pressure loss, which is governed by the pipe friction formula with a constant friction factor, so that pressure drop per unit length is proportional to velocity squared and inversely proportional to pipe diameter. Pipe costs per unit length are proportional to diameter (hint: first find the relationship between flow rate and optimum diameter).

Q.3.8(2). An axial piston swashplate type hydraulic pump (see Figure 3.16) has a fixed swash angle of 20° and n cylinders of diameter d. To accommodate the slipper pads, the centre distance between the cylinders must be a minimum of $1.25d$ and for the same reason the overall diameter D of the cylinder block must exceed the p.c.d. of the cylinders by $1.25d$. Find the number of cylinders which gives the maximum swept volume per revolution for a given value of D and the percentage decrease involved in using (a) one more cylinder and (b) one less cylinder.

Answers and notes

A.3.1. The appropriate form of C_T is given by equation 3.17, where M is the tonnage.

A.3.2. Lower: this rather surprising result is the effect of the capital cost of the engines. Remember that the b term of Section 3.3 involves costs of engines and fuel, and fuel is not burnt when not under way but engine capital charges continue.

A.3.3. (a) Lower due to engine capital cost term, higher due to fuel cost term: on balance likely to be nearly unchanged; (b) higher: since the turn-round time is proportional to tonnage, and the total tonnes per year is constant, a fixed fraction of the year is lost. Capital terms in both a- and b-type costs will be increased accordingly, but not the fuel part of b-type costs. Costs for a will be increased relatively more, so that the optimum speed rises, as with the LNG tanker.

A.3.4. $4s\sqrt{(c_1c_2\rho)}\ \dfrac{P}{V}$

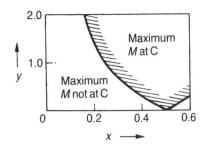

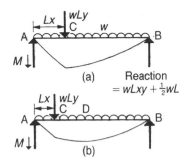

Figure 3.17   Solution to Question 3.5.

A.3.5. See Figure 3.17. Inset (a) shows the loading and bending moment (M) diagram for a case where the maximum M occurs at C, and (b) for a case where it does not. When the maximum does not occur at C it occurs at a turning value such as D in Figure 3.17(b) since there cannot be a cusp except at C. A cusp requires a sudden change of *slope* of M, and hence a *jump* in S, the shear force, since S is the derivative of M. A jump in S requires a point load.

The regions of x, y in which the maximum M is at the cusp at C and those in which it is at a turning value elsewhere are separated by a curve where the cusp and the turning value coincide. A turning value of M implies a zero of S, and so on the boundary curve $S = 0$ just towards the centre from C. From Figure 3.17 (a),

$$S(x+) = wLxy + \tfrac{1}{2}wL - wL(1 - x) = 0$$

on the boundary curve, giving

$$xy + x = \tfrac{1}{2}$$

A.3.6. 1.4 per cent. From equation 3.8, for a change x, y, not to alter π, the pressure drop term $x^{-1.8}y = 1$, so that if p is the small percentage increase in y, that in x must be $p/1.8 = 0.55p$.

A.3.7. 90°. The optimum diameter is proportional to the square root of the flow rate, so that the branches are $1/\sqrt{2}$ times the diameter of the main and $1/\sqrt{2}$ times the cost per unit length. The pumping costs are 0.2 times the pipe costs. The total cost is proportional to

$$AD + 1/\sqrt{2}\,DB + 1/\sqrt{2}DC = AD + \sqrt{2}DC$$

which is readily minimised. Indeed, by sketching two close trial positions of D it can be seen at once where the minimum occurs.

A.3.8. Seven cylinders (a) 0.9 per cent (b) 0.6 per cent.

4 Insight

4.1 Introduction

Perhaps the most important single prerequisite for good solutions to design problems is insight (see Section 1.4). From numerous instances [2] it may be inferred that insight frequently develops by large steps—it 'dawns' or 'comes in a flash'—but the steps are separated by laborious stretches of mental spadework.

By the use of aids we can hope to do two things. Firstly, we may be able to reduce the amount of spadework. But sometimes even the spadework does not suggest itself, and then we may be glad of a line on which to work, laborious or not.

This chapter discusses some approaches and devices which may be effective either in developing insight more speedily, or in developing it at all in more difficult cases (e.g. Section 4.4).

4.2 Rough calculations

A good general principle is to quantify whenever possible. It is easy to waste much time on qualitative studies of matters which might be clarified by a few lines of arithmetic. Designers should cultivate the habit of making very rough quick calculations whenever the need arises, and they should not be discouraged too easily by superficial difficulties from doing so.

It was Chesterton who said that if a thing is worth doing it is worth doing badly, an aphorism which is very true of engineering calculations. In the early stages of a design there is often little need of accuracy, and it is wasteful to make lengthy computations or to have recourse to a computer when a few minutes with a calculator will give an adequate answer. Sometimes the order of magnitude alone is enough to show that an effect can be safely ignored or that a line of thought must be abandoned, as in the case of the ice-covered methane tanker (Section 1.4). An accurate calculation done at an early stage will usually have to be repeated towards the end with different data, and is only justified if it is vitally concerned with the feasibility of the scheme. Even then, it may often be preceded with advantage by several 'rounds' of quicker, rougher calculations which may help in 'homing in' on a suitable set of trial parameters. The great Italian architect and civil engineer Nervi had this to say of rough calculations:

> The core of the problem, then, is how to develop in students a static sense, the indispensable basis of intuition of structural imagination, and how to give

them a mastery of rapid, approximate calculations for purposes of orientation
. . . The substantial difference between teaching statics to architects and to
engineers lies in the fact that architects must possess such an understanding
and mastery of the static-constructional field as to be able to *create and
approximately dimension* new structural architectural solutions, while it is
sufficient for engineers to have such a knowledge of mathematical theories as
to enable them to analyse and dimension exactly the various parts of an
already defined structure.[47]

Nervi is writing of a field in which architects are the conceptual designers and
engineers are the stressmen and detailers, whereas in other branches of
engineering (and some civil engineering too) the designers are engineers. In
general, it is even more important to engineers than to architects to possess the
'static sense' of which Nervi writes, and in the case of mechanical engineers a
dynamic sense must be added to it. The central point is the importance of 'rapid
approximate calculations', of which Nervi says, 'For more than thirty years my
calculations for orientation purposes have never gone beyond the basic arithmetic
operations.'

   To take a specific case, finite element methods (FEM) are used for most stressing
of sufficient importance, but they can only find the stresses in a defined structure
and design by their use is usually too slow and expensive. They cannot provide the
sort of insight which is given by the simple equation 4.1 in the example which
follows. In general, the designer will have to define the structure which is to be
stressed by FEM, and if his guesses are poor, a large number of expensive and time-
consuming iterations will be needed. The ideal is a designer who 'guesses' so well
that his first try is good enough, or nearly so. He will often use a computer in his
rough calculations, but the methods will be simple and the data input small, so that
a large number of iterations may be made quickly. The second case below, the box
girder, is one in which these calculations are particularly short and simple.

   The four examples which follow have been chosen to illustrate different aspects.
The first gives an algebraic approximation from which the effect of various changes
are easily seen while the second is a typical short numerical example. The third and
fourth show how insight of a lucid kind can be developed by reasoning and simple
sums.

(1) Diaphragm clamp

A slide of a machine tool has to be clamped hydraulically by means of a circular
diaphragm as shown in Figure 4.1. Suitable values for dimensions *t* and *l* are to be

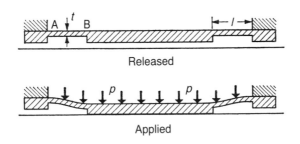

Figure 4.1   Diaphragm clamp.

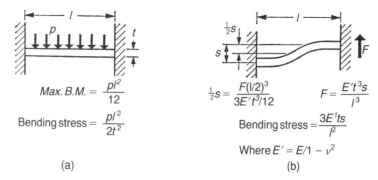

$$\text{Max. B.M.} = \frac{pl^2}{12}$$

$$\text{Bending stress} = \frac{pl^2}{2t^2}$$

(a)

$$\tfrac{1}{2}s = \frac{F(l/2)^3}{3E't^3/12}$$

$$F = \frac{E't^3s}{l^3}$$

$$\text{Bending stress} = \frac{3E'ts}{l^2}$$

Where $E' = E/1 - v^2$

(b)

Figure 4.2   Stresses in clamp.

determined, given the operating gauge pressure p, the radius r, the maximum travel s, and the allowable stress f.

To design the diaphragm using the stress analysis for a circular plate would be tedious, but for practical purposes we can regard the thinned zone as a beam, subject to two forms of loading, that due to the imposed load, the pressure p (Figure 4.2(a)), and that due to the imposed deflection s (Figure 4.2(b)). The only account we need take of the great width of the beam in the circumferential direction is that circumferential strains will be suppressed (i.e. the beam approximates to a condition of plane strain rather than plane stress), and this means that we must substitute

$$E' = \frac{E}{1 - v^2}$$

where v is Poisson's ratio, for E in the usual beam deflection formulae.

It is easily found that the maximum tensile stress, σ_M, which occurs on the oil side at A, is given approximately by

$$\sigma_M = \frac{pl^2}{2t^2} + \frac{3E'ts}{l^2} \tag{4.1}$$

and if r is a few times greater than l, the accuracy will be adequate for all useful purposes. It is much easier to see from this expression the effect of various changes, say, in s and t, than it would be with a more exact form.

(2) Design of a box girder

It is frequently necessary to design a section for a box girder to carry a given bending moment, M, and shear force, F, without exceeding some allowable stress, f. The analytical difficulties are trivial but nevertheless students have trouble.

It should first be recognised that the box or hollow rectangular section, like the I-beam, is a rational design with considerable separation of function. The bending moment is taken by the flanges (Figure 4.3) with a little help from the webs, and the shear is taken by the webs, with virtually no help from the flanges. Moreover, except with very short deep beams, the crucial function is resisting bending, not shear.

We start by assuming a depth, d, for the section (Figure 4.3). If the webs provided no bending strength, then the bending would be resisted by a tension, T, in one flange and a compressive force, $-T$, in the other, where

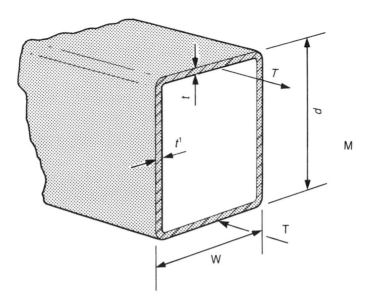

Figure 4.3   Design of a box girder.

$$T = \frac{M}{df}$$

and hence the cross-sectional area, A, of each flange would have to be

$$A = T/f = \frac{M}{d}$$

Thus, once a value of d has been assigned, the flange area A is known. We can now choose a thickness, t, and hence find the width of section w, since

$$w = \frac{A}{t}$$

This value of w ignores the contribution of the webs to resisting bending: this is easily allowed for by noting that the square of the radius of gyration (k^2) for the flange material will be $d^2/4$, while that for webs will be only $d^2/12$, i.e. unit area of web cross-section will contribute only one-third as much to bending as unit area of flange cross-section. Thus if the thickness of the webs is t', each flange is assisted by an effective web area of

$$\frac{1}{3} \times 2 \times \frac{dt'}{2} = \frac{dt'}{3}$$

and w becomes

$$w = \frac{1}{t}\left\{A - \frac{dt'}{3}\right\}$$

As a simple example, suppose $M = 2\,kNm$, $F = 10\,kN$, f is $200\,N/mm^2$, and start with $d = 80\,mm$.
   Then

$$A = \frac{2 \times 10^6}{200 \times 80}\,mm^2 = 125\,mm^2$$

Now suppose we take $t = 1.6$ mm, $w \simeq 80$ mm. The web area contributing to each flange, if we retain 1.6 mm for the web thickness t', is roughly $80 \times 1.6 = 128$ mm^2, and $\frac{1}{3}$ of this, the effective contribution, is 43 mm^2. Hence w reduces to $(125 - 43)/1.6 \simeq 51$ mm. The shear load is nearly uniformly carried by the whole web area, giving a mean shear stress

$$\tau = \frac{10 \times 10^3 \, \text{N}}{2 \times 80 \times 1.6} = 39 \, \text{N/mm}^2$$

against an allowable value (with an allowable *direct* stress of 200 N mm^2) of perhaps 100 N/mm^2. It remains to check for the stability of the webs, the effects of manufacturing methods and so on. This is the sort of simple sum of which Nervi wrote.

The remaining examples are both taken from one narrow field, that of rotational stresses in disc-like structures, which has already been discussed in Chapter 3.

(3) Turbine cooling-air shroud

Figure 4.4 shows part of a gas turbine, with a shroud disc 1 which serves to lead cooling air up the forward face of the blade disc 2. At the point A at radius r the rim of the shroud rests against the disc. The pressure of the cooling air is only slightly greater than that on the other side of the shroud. It is required to know the relative movement of the disc and shroud at the point A.

A complete stress analysis of the shroud disc is an involved calculation. It is a simple matter, however, to find an approximate solution. A thin ring of metal of radius r rotating by itself would have a stress in it of ρv^2 where ρ is the density and v the peripheral velocity (see Section 3.12). The strain in it would be

$$\frac{\rho v^2}{E}$$

and the radial growth would be

$$\frac{\rho v^2 r}{E}$$

If such a ring forms the rim of a thin *flat* disc, it cannot move radially outwards this much because the flat disc can only be stretched radially by stretching the web. But a conical disc can be stretched radially much more easily by *bending* of the web as in Figure 4.4(b) and hence much lower forces will suffice. The radial stiffness of the rim of the cone is orders lower than that of the flat disc. Thus we may regard the thin ring at A as virtually unsupported radially, so that the circumferential stress in it and the radial growth are given sufficiently accurately by the expressions above. Also since this relatively free radial growth is achieved by bending the web, it must occur as the radial component of a motion normal to the generators of the cone, as in Figure 4.4(a) and (b). Thus the axial separation of the shroud from the disc is CD in Figure 4.4(a), and

$$\text{CD} = \frac{\rho v^2 r}{E} \cot \theta$$

The writer scored a small success when he first made this calculation in his head, just before an experimental determination was begun, the result of which agreed very closely with his estimate. A second quick estimate is also needed to be sure that it is reasonable to do the first, but this will not be given here.

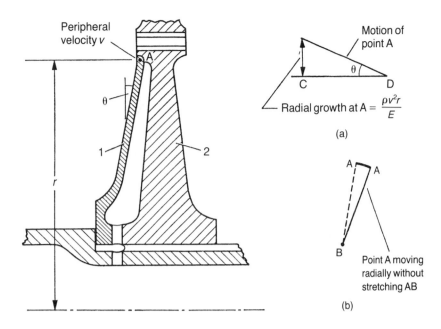

Figure 4.4   Cooling air shroud.

(4) Thermal stresses in discs

Consider a turbine disc, such as those discussed in Section 3.11 or the one in Figure 4.4, with a substantial temperature difference, say 100°C, between hub and rim. Let us try to estimate, very roughly, the resultant thermal stresses. If the material is steel with a Young's modulus of 200 kN mm^{-2} and a mean coefficient of thermal expansion of 12×10^{-6} per °C over the range concerned, then the thermal strain caused by 100°C temperature change is 1.2×10^{-3}, which could be produced by a tensile stress of 240 N mm^{-2}. The outer, hotter, part of the disc will try to grow in diameter, tending to stretch the inner part to fit it. The resisting pull of the inner part will decrease the expansion of the outer part below the 1.2×10^{-3} strain it would adopt if unrestrained. The mismatching thermal strain of the rim relative to the hub will be accommodated in three ways:

(a) by pulling in the rim, which will be in *compression* circumferentially,
(b) by stretching the web, radially, and
(c) by stretching the hub, which will be in *tension* circumferentially.

If only a narrow rim of the disc is hotter than the rest, as in the temperature distribution I of Figure 4.5, then it is clear that nearly all the mismatch will be taken up in way (a), since the amount the small hot portion will be able to stretch the remainder will be negligible. The rim will thus be subject to a compressive stress of almost 240 N mm^{-2}.

In the case of a temperature distribution such as II in Figure 4.5, we may regard the differential thermal strain as equivalent to an effective stress of 240 N mm^{-2}, of which, say, 130 N mm^{-2} appears as a compressive stress at the outside, 30 N mm^{-2} is accounted for by tensile stress at the inside, and the remainder is accounted for by radial stretching of the web. Because the circumferential length of the hub is less

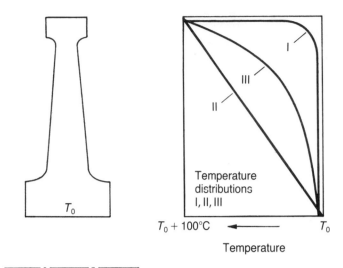

Figure 4.5   Thermal stresses in discs.

than that of the rim, the actual tensile stress in it will be much higher than 30 N mm^{-2}. The allocation of the effective stress between rim and hub must be done in inverse ratio to the areas of the sections, in order to satisfy conditions of equilibrium. A similar argument determines the allocation to the stretching of the web.

Of course, this kind of argument gives only a rough picture of the sort of stresses to be expected in the disc, but it enables us to develop a feel for the problem, so we can guess very well what will happen to the stresses if the temperature distribution changes, say, to III in Figure 4.5, or we make the rim lighter in section, and so forth. Anyone who doubts the respectability of the 'quick sum' should consider the following passage.

> Let it be assumed that the typical night-bomber is a metal winged craft well bonded throughout, with a span of the order of 25 metres. The wing structure is, to a first approximation, a linear oscillator with a fundamental wavelength of 50 metres and a low ohmic resistance. Suppose a ground emitting station to be set up with a simple horizon half-wave linear oscillator perpendicular to the line of approach of the aircraft and 18 metres above ground. Then a craft flying at a height of 6 km and at 6 km horizontal distance would be acted on by a resultant field of about 14 millivolts per metre, which would produce in the wing an oscillatory current of about $1\frac{1}{2}$ milliamperes per ampere in sending aerial. The re-radiated or 'reflected' field returned to the vicinity of the sending aerial would be about 20 microvolts per metre per ampere in sending aerial.

This passage occurs, with others like it, in a historic paper on the detection and location of aircraft by radio methods—radar, in fact—presented to the Commitee for the Scientific Survey of Air Defence on 27 February 1935.

Another historic 'quick sum' is given by Watt in a letter written to Boulton about the threat to their engine from a turbine. He deduced that the turbine blades would have to move at very high speeds to be efficient and wrote [15]:

In short without god makes it possible for things to move 1000 ft pr″ it can not do much harm.

This curt dismissal of the steam turbine as a practical possibility is an entirely adequate summary of the situation then and would merely have lost force by elaboration. It does not even matter that it would have been enough to be able to make things to move 400 ft pr″.

4.3 Optimisation of compressor shaft diameter

The problem here treated will be taken as an example of a rough calculation though it is also an optimisation (Chapter 3) and a differential study (Section 3.6). It was seen in Section 2.3 that the shaft of an axial flow compressor rotor of shaft-and-disc construction has a structural function, which demands a section of given bending stiffness per unit length EI, or, if the material is given, a fixed second moment of area of cross-section I.

If the shaft is tubular with inner and outer radii r_1 and r_0, respectively (Figure 4.6),

$$I = \frac{\pi}{4}(r_0{}^4 - r_1{}^4) \tag{4.2}$$

If r_0 is increased while keeping I fixed, the cross-sectional area

$$A = \pi(r_0{}^2 - r_1{}^2) \tag{4.3}$$

decreases and the shaft becomes lighter. However, the holes in the discs become larger and the hubs heavier. In an aircraft, turbine weight is important, so that we seek the value of r_0 that gives the minimum combined weight of shaft and hubs. In what follows it will be assumed that shaft and discs are of the same density, so that the problem becomes one of minimising volume. Consider the length of shaft p occupied by one disc of web thickness t.

Now it was shown in Section 3.12 that in an elastically designed disc the cross-sectional area of the hub reinforcement A_h (shown shaded in Figure 4.6) is given by

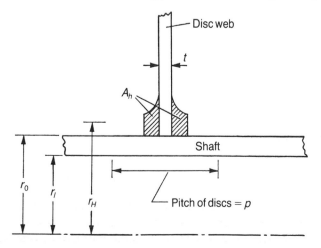

Figure 4.6   Optimum diameter shaft.

$$A_h = \frac{r_0 t}{1 - \nu} \tag{4.4}$$

where ν is Poisson's ratio. The volume of the hub reinforcement V_h is $2\pi r_h A_h$, where r_h is the mean radius of the reinforcement (Figure 4.6) and will be fractionally greater than r_0, say, $1.05 r_0$, so that, if ν is 0.3,

$$V_h = 2\pi \times 1.05 r_0 \times \frac{r_0 t}{0.7} = 3\pi r_0^2 t \tag{4.5}$$

But the volume of web cut out in making the hole is $\pi r_0^2 t$, so that the net volume penalty over a disc with no central hole is $2\pi r_0^2 t$. We might say that the weight of a hole is twice the weight of material cut out.

The additional volume associated with the hub is $2\pi r_0^2 t$ and the disc occupies an axial length p (Figure 4.6). The additional hub volume per unit length of shaft

$$\frac{2\pi r_0^2 t}{p}$$

is a kind of effective or mean cross-sectional area which may be added to the shaft cross-sectional area. (The thickness t of the webs and the pitch p of the discs will vary through the compressor, but an average t and an average p will involve no error.) We require to minimise this combined cross-sectional area A_T subject to a constant value of I. This we do by minimising $A_T + \lambda I$, subject to no restriction, by varying r_0 and r_I (see Section 3.10) where

$$A_T = \frac{2\pi r_0^2 t}{p} + \pi(r_0^2 - r_I^2) \tag{4.6}$$

We can just as well minimise

$$U = \frac{4}{\pi}(A_T + \lambda I) = \frac{8 r_0^2 t}{p} + 4(r_0^2 - r_I^2) + \lambda(r_0^4 - r_I^4) \tag{4.7}$$

and we can differentiate with respect to r_0^2 and r_I^2 rather than r_0 and r_I

$$\left.\begin{array}{l} \dfrac{\partial U}{\partial(r_0^2)} = \dfrac{8t}{p} + 4 + 2\lambda r_0^2 = 0 \\[3mm] \dfrac{\partial U}{\partial(r_I^2)} = -4 - 2\lambda r_I^2 = 0 \end{array}\right\} \quad \text{at the minimum of } U \tag{4.8}$$

Eliminating λ, $\dfrac{r_0^2}{r_I^2} = \dfrac{2t}{p} + 1$ and we find

$$r_I^4 = \frac{\dfrac{4}{\pi} I \left(\dfrac{p}{t}\right)^2}{4\left(1 + \dfrac{p}{t}\right)}, \quad r_0^4 = \frac{4}{\pi} I + r_I^4$$

These results are plotted in non-dimensional form (such that $r_0 = 1$ for a solid shaft) in Figure 4.7. As might be expected, the highest values of p/t (i.e. thin discs like compressor discs) give thin-walled optimum shafts, while thick discs give thick-walled shafts. The argument has been simplified by ignoring the influence the

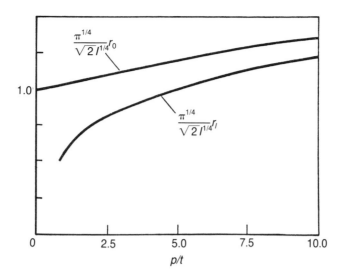

Figure 4.7 Variation of inner and outer radii of minimum weight shaft with pitch/web-thickness (p/t) ratio.

variation of the mass of the rotor has on the value of I required, an effect which causes only slight complication. Other slight modifications are needed to deal with the effect of splines or other details which come between the shaft and the hub, and make r_0 different from the disc bore radius.

An alternative to this procedure, which the writer has known to be used, is to do several almost complete designs of compressor rotor with different shaft diameters, and calculate their weights. This is very expensive, takes much longer and is also much more likely to yield a misleading result. The designers working on the various schemes and the stressmen are quite likely to introduce errors, through small differences which are not associated with the changes in shaft size, through rounding small dimensions to different convenient figures, through differences in taking scaled measurements, or through insufficiently standardised design or stressing methods for the discs etc. The differential study (Section 3.6) we have made is cheap, quick, and reliable, because it deals only with the change we are interested in.

4.4   The optimum virtual shaft: a digression

Before leaving this example, let us look at a more advanced case. So far we have regarded I, r_0 and r_I as constant throughout the length of the shaft. But the requisite stiffness could be achieved with less material by making I greater at the mid-length of the shaft and less towards the end. While such a variable cross-section scheme would be expensive to manufacture in a discs-and-shaft construction, it is eminently feasible in some virtual shaft designs. In such an arrangement r_0 and r_I are functions of x, the distance along the shaft (Figure 4.8) and I is given in terms of r_0 and r_I by equation 4.2, and is also a function of x. We now need to choose these functions to give a rotor of minimum weight, subject to the condition that the

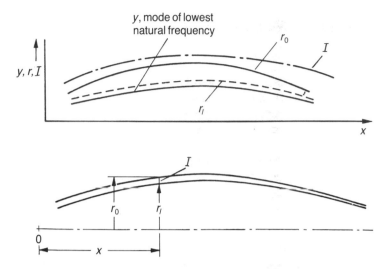

Figure 4.8   Optimum virtual shaft.

lowest natural frequency of transverse vibrations of the shaft shall have a prescribed minimum value (see Section 2.3). This is a brachistochrone problem (Section 3.10) subject to a very complicated constraint. It is too advanced for this book, so mention will only be made of the salient features, which are particularly interesting.

Once we know the value I at any section, we can find the values of r_0 and r_1 which will give this I for a minimum weight penalty by the method already described. We specify the shape or mode in which the shaft vibrates, and from this and the required minimum natural frequency decide the value of I corresponding to each value of x. Finally, the mode of vibration is determined from the 'shape' of r_1 as a function of x. From a guessed shape of r_1 (see Figure 4.8) we determine a corresponding mode y of vibration and hence a shape of I. The actual values of I are then fixed to give the required natural frequency, and r_0 and r_1 determined from I by the method already described. This gives in general a different shape of r_1, so the process is repeated until it converges, i.e. until it keeps on giving the same shape of r_1 every time.

There are two things to note especially. Firstly, we decide on the shape (mode) and frequency of vibration we want, and hence the required stiffnesses. This is not only a *design*, as against an *analytical*, process; it is much easier than finding the mode and frequency given the stiffnesses. In just the same way we saw in Section 3.11 that it is much easier to design a disc to have a given (purposeful) stress distribution than to find the stresses in a given (purposeless) design, the typical analytical problem.

Secondly, the reason for the final design being one of minimum weight lies in the method of deriving the vibratory mode y from r_1. We choose y so that the maximum value of the strain energy of vibration of the material at r_1 exceeds the maximum value of its kinetic energy of vibration by the same amount at all x. This is an 'autonomous local optimising principle' (ALOP—see Section 3.11) and automatically produces a rotor of minimum weight.

Unlike the simple method for a uniform shaft, the writer does not know of this design process being used for an existing engine; these few notes are included as an example of the automatic design of an optimum structure.

4.5 Useful measures and concepts

Another good habit for the designer is the use of known convenient measures which help to characterise design elements, and the invention of new ones *ad hoc* to help him in particular problems. Examples are joint efficiency, solidity, aspect ratio, water equivalent, block coefficient—all these are handy measures which help the mind familiar with them to visualise or assess the object or the situation described.

Such terms are so frequently used by all engineers that it may seem unnecessary to emphasise the point. But sometimes the wrong measure is used, or no measure at all where one would be helpful. Sometimes no accepted measure exists, and then the designer should devise his own. We have already had occasion to use such an *ad hoc* measure in the suspension bridge example of Section 3.4, where the cost of the cables involved the multiplier c_c/f_c where c_c is the cost of the material per unit volume and f_c is the working stress. Then the ratio c_c/f_c may be regarded as the cost per unit of strength, and it is a useful concept in the economical design of structures [16].

A common sort of useful measure is a figure of merit for materials for use in a given role. Consider helical compression springs—the wire in them is loaded in torsion, and the maximum stresses occur in the surface, which is in a state of pure shear, i.e. the principal stresses are $(-\sigma, \sigma, 0)$. If these stresses are to be the maximum allowable according to some such basis as maximum shear stress (Figure 4.9), clearly a necessary condition if the material is to be economically used, then the working point must be point

$$C\left(-\frac{f}{2}, \frac{f}{2}, 0\right)$$

Since point C corresponds to pure shear with a shear stress $f/2$ the strain energy per unit volume is

$\frac{1}{2}$(shear stress) × (shear strain) or

$$\frac{1}{2}\left(\frac{f}{2}\right) \times \frac{f/2}{G} = \frac{f^2}{8G} = \frac{(1+v)f^2}{4E} \tag{4.9}$$

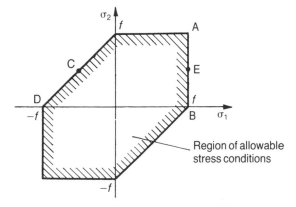

Figure 4.9 Allowable stress, maximum shear stress (Tresca) criterion.

where E = Young's modulus, G = shear modulus and v = Poisson's ratio. Expression 4.9 may be taken as a figure of merit for a material for use in a spring so long as the volume used is of overriding importance. If cost of material is most important, the figure of merit becomes

$$\frac{(1+v)f^2}{4Ec}$$

where c is the cost per unit volume.

We might consider whether point C is the best working point, and this will lead to another useful measure. It is readily shown that at points A and B of Figure 4.9 the strain energy per unit volume u is

$$\frac{(1-v)f^2}{E} \quad \text{and} \quad \frac{f^2}{2E}$$

respectively, so that if $v = 0.3$, a common value for metals, the relative values of the u's are

A	B	C
2.15	1.54	1.00

It is scarcely possible to work the material of a one-phase spring at point A, but there is no great difficulty in working it at B, e.g. in a multiple leaf spring. However, what matters is not the maximum strain energy per unit volume at the most highly stressed point, but the average strain energy throughout the material. Defining the *efficiency of use* of the material η_u as the ratio of the average to the maximum strain energy per unit volume, for the cylinder in torsion η_u is $\frac{1}{2}$, whereas for the rectangular cross-section beam in bending it is only $\frac{1}{3}$. This brings the multiple leaf spring and the helical compression spring about level as regards average strain energy per unit volume.

Much higher values of η_u are possible in some forms of spring. A hollow torsion bar will give η_us of 0.80 or so, and a particularly high average strain energy may be had in the ring spring (Figure 4.10) resulting from a high η_u combined with working point B and its compressive equivalent. Because of the sliding friction between the rings, the ring spring has a hysteretic behaviour which limits its application but can be advantageous.

On the other hand, a simple uniform-section single leaf spring has an η_u of only 1/9; not only is the average strain energy only one-third of the maximum over any section, but the maximum itself varies from section to section as the square of the

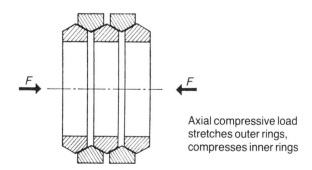

Axial compressive load
stretches outer rings,
compresses inner rings

Figure 4.10   Ring spring.

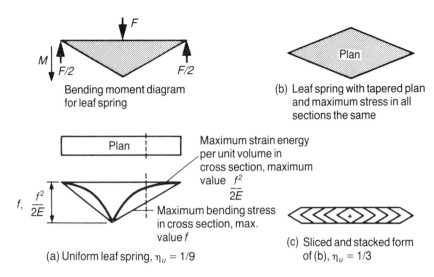

Figure 4.11   Material use in leaf spring.

bending moment. This can be overcome by narrowing the breadth of the spring in proportion to the bending moment as in Figure 4.11(b) so that the maximum stress in all sections is the same. The result is a rhombic form which, in vehicle leaf springs, is effectively cut into strips which are stacked one above another as in Figure 4.11(c). Notice that vehicle springs do not work quite like the rhombic spring—apart from anything else friction forces between them prevent them working entirely as independent beams—but the efficiency of use of material is nearer the ideal 1/3 than the uneconomical 1/9 of the spring of Figure 4.11(a).

Suppose a tank of liquid nitrogen is to be suspended within an outer container; what is an appropriate figure of merit for the material of the suspension if it is required to keep the heat leakage into the tank a minimum? The immediate reaction might be to think of conventional insulating materials, but the figure of merit is f/k, where k is the thermal conductivity and f the allowable stress; if the material is stronger, we can use a thinner section which will conduct less heat. Such materials as stainless steel and Inconel will rate well on this basis.

Thermodynamic efficiency

One last and rather long example of useful measures will be given. A ritual piece of information which is instilled into engineering students is the definition of the coefficient of performance of a refrigerator as the heat extracted from the thing cooled divided by the work expended. However, heat extracted is a very poor measure of the task performed since it does not take account of the temperature difference through which that heat must be 'pumped'. A performance ratio of 10 in a desalination plant would be poor, and one of 0.01 in a helium liquefier remarkably good.

A better measure of the performance is the thermodynamic efficiency η, defined as the ratio of the work required by a reversible machine performing the same task to the actual work. If we call the work required by the reversible machine the 'ideal work',

$$\eta = \frac{\text{ideal work}}{\text{actual work}} \text{ for a refrigerator or a heat pump,} \qquad (4.10)$$

and, by extension, for an engine

$$\eta = \frac{\text{actual work}}{\text{ideal work}} \qquad (4.11)$$

where the ideal work is that obtainable from a reversible machine using the same heat supply or the same fuel.

The ideal work required to effect a change in one unit of a substance (e.g. liquefy nitrogen) in a steady flow process, say from state 1 to state 2, is given by

$$b_2 - b_1 = \Delta b$$

where $b = h - T_0 s$, h is specific enthalpy, s is specific entropy and T_0 is the temperature at which heat is rejected, which is generally the ambient temperature. The function is called the availability (strictly the steady flow availability function), and the increase in availability Δb is a *measure* of the thermodynamic *task* to be performed.

Suppose now we imagine the change effected in two stages, from state 1 to state M and then from state M to state 2, the actual works required being W_{1M} and W_{M2}. Then there is an efficiency of each stage, and we can put, say

$$\eta_{1M} = \frac{b_M - b_1}{W_{1M}} \text{ and } \eta_{M2} = \frac{b_2 - b_M}{W_{M2}}$$

Clearly the higher the ηs of the individual steps, the higher the overall η and the lower the work consumption of the process.

With these two useful concepts, thermodynamic efficiency and availability, at our disposal, let us look at the problem of liquefying nitrogen initially at room temperature and pressure, a task involving an increase of availability of, say, 100 units. Then it seems a reasonable plan to look for a practical process, *any* practical process, by which we can increase the availability of the nitrogen very substantially and as efficiently as possible. There are two such processes

    (a) raising the nitrogen to a high temperature,
    (b) compressing it highly.

Now since we seek to liquefy the gas (a) seems rather unpromising, which leaves (b). We can compress the nitrogen with a thermodynamic efficiency of about 0.80, and raise its b by about 60 units, which takes

$$\frac{60}{0.80} = 75 \text{ units of work}$$

to drive the compressor. The remaining increase of b of about 40 units can be achieved by cooling with a suitable refrigerator of one of the known sorts, which might have an efficiency of about 0.5. Thus the total units of work required are

$$75 + \frac{40}{0.5} = 155$$

giving an overall thermodynamic efficiency of

$$\frac{100}{155} = 0.65$$

If we liquefied the nitrogen entirely by cooling with a refrigerator of efficiency 0.5 we should need 200 units of work.

This deceptively simple argument reaches conclusions which are correct and might be valuable were there any need to liquefy nitrogen in large quantities, and it is a line of reasoning which is very easy to find once the two concepts on which it is based have become slightly familiar. Such concepts are valuable tools for the designer.

4.6   Bounds and limits

In the elementary theory of oil-lubricated bearings, there comes a point where it seems that the load capacity rises with the speed. As the engineer is likely to want to increase the speed of machines and the loading of their bearings at the same time, such behaviour seems very convenient and so very unlikely. We should at once look for the limit to this happy state of affairs, implicit in the parameter usually written ZN/P, where Z is the lubricant viscosity, N the speed, and P the specific loading of the bearing; if we increase N and P proportionately, the value of the parameter remains unchanged so long as there is no change in Z. But, as N and P are increased, the temperature in the vital area of the lubricating film will rise, causing a relatively very sharp fall in Z. Eventually a point is reached where the product ZN, and hence the load capacity, falls as N increases.

The engineer can perhaps be forgiven a little anthropomorphism, if he thinks of the natural order as imbued with cussedness, never giving with one hand except to take away with the other. He rarely makes an advance in one direction except at the expense of losses in another, but it is this very sort of difficulty that gives his work its challenge. An important aspect of insight into an engineering problem is a clear and balanced view of the conflicting constraints set to its solution by natural laws, available materials and the state of the art.

Consider the problem of designing a beam AB (Figure 4.12) of minimum weight, subject to a given bending moment M and shear force S, which are specified functions of x, the distance from A. The allowable stresses are $f/2$ in pure shear and f in simple tension or compression, corresponding to the maximum shear-stress criterion used elsewhere (Figure 4.9). If the depth of the beam is h and it consists of two flanges, of cross-sectional area A_F each, which take the bending moment, and a web of area A_W that takes the shear,

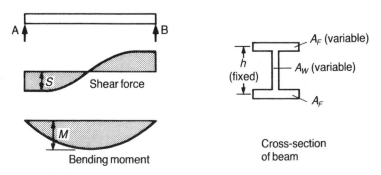

Figure 4.12   Structural task of a beam.

$$A_F = \frac{M}{fh} \text{ and } A_W \simeq \frac{2S}{f}$$

and the total volume of material

$$\int_A^B (2A_F + A_W)\, dx = \frac{2}{fh} \int_A^B M\, dx + \frac{2}{f} \int_A^B S\, dx \qquad (4.12)$$

This analysis will do fairly well as long as the beam is shallow and the structural loading coefficient is high enough and the material stiff enough for no questions of stability to arise—put another way, as long as the beam is of high solidity. But if we attempt to minimise expression 4.12 with respect to h it becomes clear at once that its validity must be limited, since the first term in it decreases indefinitely as h increases, and the second is independent of h. Expression 4.12 has no minimum, but the engineer's view of the physical world tells us there must be one; hence expression 4.12 must be an inadequate representation of a deep beam.

There must be a solution to the problem of the minimum-weight beam, and a sufficient number of particular solutions of this sort of problem are known for us to estimate the volume of material in it fairly closely. This volume we shall call 'ideal', by extension of the use of 'ideal' in thermodynamics. Normally minimum-weight structures are highly impracticable (not in the case of the disc; see Section 3.11) but practicable designs can be found which are not very much heavier. The ratio of the volume of material in a practical design to the ideal volume is a good measure of its excellence as regards economical use of material. Moreover, a study of the 'ideal' form will provide a great deal of help in designing a good practical form [17].

The most famous limitations in engineering science are the first and second laws of thermodynamics; the statements of the second law as an embargo on perpetual motions of the second kind and as Carnot's theorem are indeed expressly limitations of possible engineering devices. The scope of the second law is not always realised; consider the following proposal to enable soldiers to see better at night.

> Make field glasses with object glasses ten times the diameter of the pupil of the eye (about 5 mm) and a magnification of five times. The object glasses will collect 100 times as much light as the naked eye (say 90 times allowing for absorption and reflection by the glass) but only produce an image on the retina twenty-five times the area, and so $90/25 = 3.6$ times as bright.

If it were possible to construct such an optical system, an image of the Sun could be formed which would be brighter than the Sun itself and, with suitable insulation of the area on which it was formed, would become hotter than the Sun. We should then have heat flowing of its own accord from a colder to a hotter body, which is impossible by the second law, so the device will not work (in the way claimed, anyway). What happens is that only roughly a quarter of the light leaving the eyepiece enters the pupil. We could prove this by geometrical optics, but the proof by the second law of thermodynamics is immediate and rigorous.

In the past, inventors pursued the will of the wisp of the perpetual motion of the first kind, and modern favourites are machines that disobey Newton's third law of motion that to every action there is an equal and opposite reaction. The obvious form is the reactionless drive (or, possibly, the anti-gravity device) but another variety is the variable-speed gear based on a differential. One shaft of a differential is driven, another is the output, and the third is allowed to run away, as it were, but only in a controlled fashion. This is a failure of insight, an inability to recognise that the torque in one shaft determines those in the others and so a wide speed range can be achieved only with a large power flow in the third shaft. To feed this power into the output we should need a variable-speed gear, which is what we started out to design.

Nevertheless, this is not a fruitless train of thought. Suppose an automatic transmission for a car needs a power throughput of 50 kW and a ratio of top to bottom gear, what we shall call the range of ratio, R, of 4. If we use a hydrostatic transmission, the pump has to be capable of absorbing 50 kW, which is a fair measure of its size and cost. If at maximum power in bottom gear the hydraulic motor speed is ω and the torque it develops is T, then it has also to run at 4ω at maximum power in top. Since $\omega T = 50$ kW, the product of the maximum torque T and the maximum speed 4ω is 200 kW. This 'power', called the 'corner power,' is another useful concept. Although the true power never exceeds 50 kW it is nearly as exacting to produce a motor with a product of maximum torque and speed of 200 kW as one with a true power of 200 kW. The corner power is a good rough measure of the hydraulic task and hence the amount of machinery and its cost. Thus the whole transmission requires about 250 kW-worth of hydraulics.

Let us take the third shaft of our original differential and connect it via a hydraulic variable-speed gear (v.s.g.) to the output shaft (Figure 4.13). We now have an arrangement in which the power is divided in the differential, flowing through two branches A and B to be recombined in the output shaft. As only a part of the power flows through the branch with the hydraulic elements, the corner power is reduced. For the same reason, the variation in the overall ratio of the whole system is less than that in the hydraulic 'branch', but the range of ratio of the hydraulic variable-speed gear is infinite and so a reduction can be tolerated. In fact, the whole arrangement may be regarded as 'trading off' the excess range of ratio available in the hydrostatic elements for increased throughput. The question is, just how low can we make the required corner power of the hydrostatic elements in this way? What is the 'conversion rate', as it were, for exchanging excess range of ratio for higher throughput?

The answer in the case of the car transmission is disappointing: without reversal of power flow in the A branch (which is awkward and inefficient), we cannot reduce the corner power of the hydraulic elements below 180 kW, which is not enough to make the scheme viable. On the other hand, in an application which only needed a range of ratios R of 1.5, such a scheme could reduce the corner power requirement by two-thirds, a very worthwhile saving. The split drive can reduce the total corner power in the ratio $(R - 1)/R$ at most; this restriction applies to 'input and output summer' arrangements, to the type reported in reference 18, and to much more complicated schemes also.

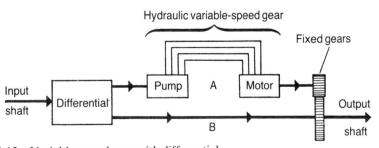

Figure 4.13   Variable-speed gear with differential.

Such split drive transmissions have found application in feed pumps, constant-speed drives for aircraft generators, and a few other fields. They are of little use in vehicles because the extent to which the corner power can be reduced is small when R is large. However, what can be done is to make a split drive which has two separate regimes of operation, upper and lower, as is done in the British Leyland continuously-variable transmission[49]. At the transition between the lower and upper regimes, the drive is in both at once, which is possible because the ratio is exactly the same in each. The lower regime goes from reverse, through a geared neutral into the low forward gears, and then the upper regime takes over and goes up to top gear. In this case, the range of ratio for the transmission, because it includes negative values in reverse, is larger than the range of ratio of the continuously-variable transmission, which only gives positive ratios, and the saving in corner power in using a split drive with two regimes is about half.

The British Leyland transmission uses a continuously variable element of the traction or 'friction drive' type, not a hydraulic one, and it seems quite likely that this type of drive, with either traction or compression belt variable elements, will be used generally in vehicles in the future.

4.7   Scale effects

In many problems a useful way of acquiring insight is to study the effects of scaling. If all the dimensions of a pressure vessel are increased by the same scale factor (including wall thicknesses) then the stresses are unaltered. If all the dimensions of a gravity-loaded structure such as a suspension bridge are scaled up, the stresses increase in the same proportion.

Engines

Consider an internal-combustion engine, which runs with angular velocity ω and has a crank throw L. Then the centripetal acceleration of the crank pin is $L\omega^2$, and all the accelerations of the moving parts are proportional to $L\omega^2$. If we take L as a typical dimension, then the mass of the connecting rod, for example, is proportional to L^3 and so the inertia load it exerts on the bearing is proportional to $L^4\omega^2$. The bearing area is proportional to L^2, so that the specific loading of the bearing is proportional to $L^2\omega^2$. In a similar way, the inertia stresses in the connecting rod are proportional to $L^2\omega^2$. If we scale the engine, then we need to keep $L\omega$ nearly constant if the inertia stresses are to remain acceptable.

Let us see now how this affects the power/weight ratio of the engine. The swept volume and hence the work done per stroke increases as L^3, but ω and hence the number of strokes in unit time is proportional to $1/L$, so that on balance the power is only proportional to L^2 while the weight is proportional to L^3. This is sometimes called the '$1\frac{1}{2}$ power law', because the weight increases as the power to the index 1.5.

One way round the $1\frac{1}{2}$ power law is to increase the swept volume by increasing the number of cylinders without making them bigger. Further, since the power for a fixed maximum speed (fixed ωL) depends only on piston area, the weight per unit power is inversely proportional to the stroke. Logically, then, to make a light engine, we make the stroke as short as may be relative to the bore ('over-square') and use a large number of small cylinders—a philosophy embodied in the original BRM engine.

Turbines

The case of a turbine is similar. The mass flow rate of fluid, and hence the power, is proportional to the square of the diameter. If all the dimensions are scaled up equally, the power is proportional to L^2 and the weight to L^3 and we have the same $1\frac{1}{2}$ power effect. But we have rather more freedom than this. Why need we scale up the axial dimensions by the same factor as the diametral ones, indeed, why need we scale them up at all?

The answer lies in the bending stresses in the blading. Imagine that we scale all the transverse dimensions (including the diameter and all the blade lengths) by one factor and all the axial dimensions (including the blade chords, disc thicknesses, overall length etc.) by another, and let D and L be typical transverse and axial dimensions. The weight is proportional to D^2L. The projected area of a blade surface is proportional to DL, and since the pressure difference over it is fixed, the load on it is proportional to DL also. The bending moment at the fixed end of the blade is proportional to D^2L, and the section modulus there is proportional to L^3. The bending stress in the blade is proportional to D^2/L^2 and so remains unaltered if we scale the axial dimensions equally with the transverse ones. Thus the $1\frac{1}{2}$ power law will be followed as long as the axial dimensions are determined by stressing, and the ratio of weight to power will be proportional to the square root of the power, as in the curve EBC of Figure 4.14.

But suppose we scale down a turbine, say, represented by point C in Figure 4.14. The axial dimensions (the blade chords included) are scaled down proportionately with the transverse ones until a point is reached where it becomes impracticable to reduce the smallest ones any more. From this point (B in Figure 4.14) onwards, only some of the axial lengths can be scaled down, and so the curve drifts away from the parabola EBC. By the time even the largest blade has its chord determined, not

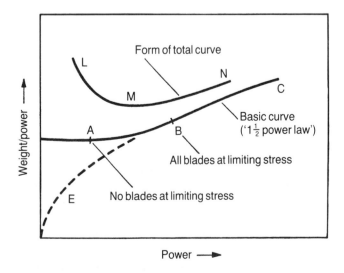

Figure 4.14   Specific weight of aircraft gas turbines.

by stress, but by the smallest size it is simple to make, no further axial scaling is possible and the weight/power ratio on this simple analysis will remain constant.

Of course, these considerations only apply to the blading and other parts governed by the blading, like discs. The casings where they are determined in thickness by pressure stresses will thin proportionately with the diameter and so will march with the blades. In small sizes the thickness may be nearly independent of diameter, and so the casing weight/power ratio will eventually increase with decreasing power. At sizes above point B in Figure 4.14, the parameter measuring the difficulty of providing sideways stiffness in rotors, the x axis in Figures 2.2–2.6, will remain constant, but below B it will increase because the length decreases less rapidly than the diameter. Proportionately more material will be needed for transverse stiffness, and this effect also will curl the curve upwards. Many accessories (and many details such as bearing assemblies) will not decrease in weight proportionately with the power. When all these are taken into account, the form of curve obtained is something like LMN in Figure 4.14, with a minimum at M.

Electrical machines

If the centrifugal stresses are limiting, an electrical machine behaves differently under scaling. If the flux and current densities are kept the same, the current taken and the total flux each increase as L^2. The centrifugal stresses are unaltered if the surface speed remains the same, so the angular velocity is proportional to L^{-1} and the e.m.f. is proportional to the total flux times the angular velocity, or L. The power is proportional to the e.m.f. times the current, or L^3. This simple result is slightly complicated by the field windings not following quite the same law, a matter we shall not discuss, and also by the problem of cooling, which is worth studying here.

In the scaled machine, the copper losses per unit volume are unaltered; if the lamination thickness is scaled, the iron losses per unit volume are also unchanged. The total heat evolved within the materials is proportional to L^3, but the surface available for cooling is only proportional to L^2. Very roughly, if no compensating measures were taken, the temperature rise would be in proportion to L. If it were acceptable on a small machine, it would not be so on a bigger one, and the power would have to be limited to an L^2 law. Maintaining the power proportional to L^3 depends on the provision of better cooling. At the lower end of the size range this can be done simply by increasing the local heat transfer coefficient over the areas naturally available, by forcing air past them at higher speeds, but in large machines extra area is needed, ideally, of course, sufficient to keep the total cooling surface proportional to L^3. This means a fixed passage size.

Cases involving flow

A rough parallel to this problem is found in the respiration of animals. Very small creatures are able to respire directly through the cell walls. Insects have branching systems of ventilating tubes opening directly to the outside which supply air throughout their bodies; the pumping of the air is effected by muscular movements which operate on the tubes as if on bellows. The much larger mammals have an even more elaborate system, using an intermediate fluid (blood) with its own pump (heart) as well as a large surface area in the lungs and an air pump (rib cage, diaphragm and associated muscles). It has been suggested, very plausibly, that

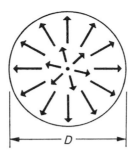

Figure 4.15   Flow of heat outwards in a rod.

there are no large insects in spite of the great success of this versatile basic design of animal because of the limitations imposed by their respiratory system. The problem here, however, is more than that of keeping the surface area proportional to L^3: it involves the work required to pump the air in and out of the system, and the space occupied by the tubes (see Question 4.9). As the insect is made larger, the amount of air flowing through unit volume of its body to reach more interior points increases in proportion to L. Similar problems arise in town traffic [19] and nuclear-reactor cores.

As a simple example of this last sort of problem, consider heat being evolved in a long rod of diameter D at a constant rate per unit volume and extracted at the surface (Figure 4.15). The total heat flow per unit length of rod is proportional to D^2, and the area of its flow path in the material to D, so that the heat flux density, and hence the temperature gradient, is proportional to D. Thus the temperature difference between the centre and the surface of the rod is proportional to D^2.

Scale effects in gearing

In Chapter 6, the design of gears will be considered, and it is convenient to study the dimensional aspects here. The loads we can apply to gear teeth are limited by bending stresses and surface stresses. If we scale a gear tooth up, the modulus of its root section increases as L^3, the arm of the tooth load about it increases as L, and so the load which can be carried increases as L^2. The torque on the gear can thus be increased as L^3.

The radius of curvature of the tooth profile is proportional to L, and so therefore is the load per unit length of facewidth that can be carried. The facewidth is proportional to L, so once again we have a load capacity proportional to L^2 and a torque capacity proportional to L^3. In this case we can scale the length L and the diameter D independently and find that the torque capacity is proportional to D^2L, which is just $4/\pi$ times the volume of the pitch cylinder. Indeed, we can write for a pinion of high quality

$$\text{allowable torque} = \frac{CD^2L}{1+k} \tag{4.13}$$

where C is a constant and k is the number of teeth in the pinion divided by the number of teeth in the gear. The writer has used $C = 300$ p.s.i. ($2\,\text{N}\,\text{mm}^{-2}$) roughly for case hardened and ground $4\frac{1}{4}$ per cent Ni-Cr steel; at very high surface speeds, where design has to be modified from what would otherwise be the optimum to avoid scuffing, C is reduced somewhat, as in rocket pump drives [20].

4.8   Dimensional analysis and scaling

The reader who has a knowledge of dimensional analysis will perhaps wonder why
it has not been used here. The reason is that scaling ideas are a more powerful and
sophisticated device, and, being capable of direct physical interpretation, are both
more easily understood and more useful for gaining insight.

Everything that can be done by formal dimensional analysis can be done by
scaling, but a very great deal which can be done by scaling cannot be done by
dimensional analysis. On the other hand, dimensional analysis is quicker, surer,
and tidier so far as it goes. We should use whichever fits the situation better (and
that sometimes means both).

The advantages of scaling are that

(1) we can tackle the problems in which there is not strict similarity, and
(2) we can use physical reasoning.

In the usual treatment of pipe flow by dimensional analysis, physical reasoning is
used when it is assumed that, for fully developed flow, the pressure gradient along
the pipe is constant. The difference between the approaches with and without
physical reasoning is brought out by the following example.

A roller of radius R, length L is pressed against the edge of a semi-infinite plate,
thickness L of the same material by a uniform force per unit length P/L. What is the
maximum stress σ_M and the halfwidth b of the contact strip?

This is a simple form of Hertz's problem. By dimensional analysis we can only
say: the parameters involved are P, R, L and E, the Young's modulus of the
material, and v, Poisson's ratio. The effect of v is nearly always small and we ignore
it. Then

$$\sigma_M = \frac{P}{RL} \, \mathrm{f}\left(\frac{R}{L}, \frac{P}{ER^2} \right) \tag{4.14}$$

where f is some unknown function of R/L, P/ER^2. This equation is obtained by
putting σ_M equal to some group with the dimensions of stress (we have chosen
P/RL, but we could have taken E) multiplied by a function f of all the independent
dimensionless groups we can form from the parameters P, R, L and E. The 'π-
theorem' tells us there are two such groups.

Now strictly speaking this is *all* dimensional analysis can tell us. But by physical
reasoning, we expect the behaviour in all sections perpendicular to the axis of the
cylinder to be the same, so that P and L will only appear as P/L, the load per unit
length, and instead of equation 4.14 we can write

$$\sigma_M = \frac{P}{RL} \, \mathrm{g}\left(\frac{P}{ERL} \right) \tag{4.15}$$

where g is an unknown function of the single parameter, P/ERL.

To proceed further, we use Figure 4.16 and scaling. If Q is a point at the edge of
the contact flat, the two surfaces there are parallel to each other after loading, while
before loading they had an angle b/R between them. The sum of the rotations of
elements at Q must therefore be b/R. Now the rotation of an element of a strained
body relative to some other element (e.g. one at R) is proportional to the strains in
that body so that σ_M/E, which is a strain, is proportional to b/R.

The total load P is proportional to the area of the contact flat times a typical
stress, or

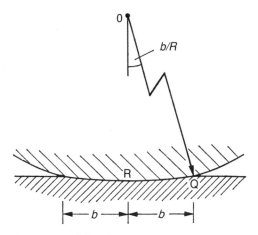

Figure 4.16   Cylinder in contact with a plane.

$$P \propto \sigma_M . bL \propto \frac{bE}{R} . bL, \; b \propto \sqrt{\left(\frac{PR}{EL}\right)}$$

$$\sigma_M \propto \frac{bE}{R} \propto \sqrt{\frac{PE}{RL}} \qquad\qquad\qquad (4.16)$$

Now this is as far as we can get, but it is a good deal further than equation 4.15. We know that

$$g\left(\frac{P}{ERL}\right) = K\left(\frac{P}{ERL}\right)^{-1/2}$$

and only the constant K is undetermined. In fact it is about 0.42 for $v = 0.3$.

This example has little to do with design, but it shows very well the power of the extended form of dimensional analysis, using physical reasoning, which is here called 'scaling'.

4.9   Proportion

There is much truth in the adage, 'if it looks right, it is right', and perhaps even more in the converse, 'if it looks wrong, it is wrong'. A good designer has the sort of feel for proportion which enables him to look at a drawing and spot the weaknesses or the inappropriate size of certain features. A simple example will show how this can be so.

Suppose a pinion meshes with a single wheel and is mounted in roller bearings at each end. If, as will usually be the case, the principal function of the bearings is to carry the reaction to the tooth load on the pinion, this consideration will fix the relative sizes of the components, so that even a slightly practised eye can recognise wrong proportions. The load-carrying capacity of the pinion tooth is proportional to the facewidth and the relative radius of curvature at its point of contact with the wheel (Section 4.8) and the radius in question is in turn roughly proportional to the pitch diameter (say, about 1/6 of it in a well-designed pinion meshing with a wheel with about three times the number of teeth).

Similarly, the load-carrying capacity of a roller bearing with a given number of rollers (say, 11) will be proportional to the facewidth (the length) of the rollers and the relative radius of curvature at their point of contact with the outer race, which is about 1/12 of the pitch diameter of the roller cage. A common assumption is that the heaviest loading on a single roller is about $5z$ of the total load, where z is the number of rollers. If $z = 11$, we can regard the load as shared among 11/5 or 2.2 rollers. If the pitch diameters of the roller cage and the pinion are about the same, and if the pinion is case-hardened and ground, so its surface strength is comparable with that of the rollers, the 4.4 effective rollers in the two bearings, having half the relative radius of curvature of the gear teeth, are suitable for carrying the tooth load on a pinion facewidth 2.2 times the roller width. Try this on the planet bearings of Figure 7.7, but allow for the smaller pitch circle radius.

4.10   Change of viewpoint

A most useful trick for increasing insight, but one rather difficult to illustrate briefly, is changing the viewpoint.

Suppose we are designing some bathroom scales; the suspension of the platform must be such that the indicated weight is the same wherever the person stands on it. When a load W is placed on the platform there is a load kW in a certain link of the mechanism; the requirement is that k should be a constant and not vary with the position of W on the platform. This is a statical condition and it can be designed for.

If we imagine the link given a small movement h along its length, then the point on the platform where W is situated moves s, say in the vertical direction. Since the mechanism is almost frictionless, we can apply the principle of work and put

$$hkW = sW, s = hk$$

Now since k is to be constant and independent of the position of W, s must also be independent of the position of W, i.e. the platform must remain parallel to its original position. This is a kinematic condition and it can also be designed for, and because mechanisms which satisfy it can be visualised relatively easily, this changed view of the problem is more likely to be fruitful. Most readers will be able to think of various parallel motions which might be adaptable to use as a platform suspension, and the easier visualisation of the kinematic property may lead to less obvious forms suggesting themselves.

The absorption refrigerator

A good example of change of viewpoint is given by the absorption refrigerator. As a reminder, Figure 4.17 shows the flow diagram of an ordinary vapour compression refrigerator. To fix ideas, the refrigerant is taken to be ammonia, which is compressed and condensed at roughly ambient temperature and the higher pressure in the condenser. The resulting liquid ammonia is passed through the throttle, yielding mostly liquid and a little vapour at a lower temperature and pressure. The liquid is then evaporated in the evaporator, extracting heat from whatever it is required to cool.

Figure 4.18 shows the flow diagram of an absorption refrigerator working on

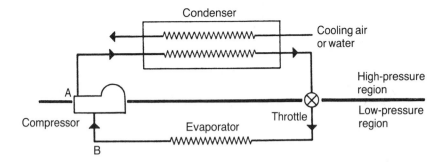

Figure 4.17   Vapour compression refrigerator

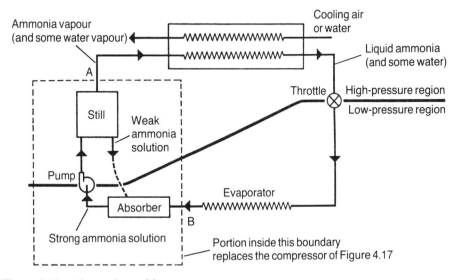

Figure 4.18   Absorption refrigerator.

ammonia and water. From A to B it is identical with the vapour-compression cycle. Instead of using a compressor, however, the ammonia vapour leaving the evaporator at B is absorbed in a weak solution of ammonia to give a strong solution of ammonia with a volume perhaps one per cent of that of the vapour. This strong solution is raised to the higher pressure by a liquid pump. Then the ammonia vapour is boiled out of the solution by the application of heat in a still, leaving a weak solution which is returned to the absorber. The available energy input comes mainly from the heat source used for the still, the liquid pumping term being nearly negligible.

Thus the absorption refrigerator can be viewed as a vapour-compression refrigerator in which the compressor is replaced by an absorber, liquid pump and still. This is already a fairly abstract view, and a helpful one. But we can do a great deal better.

A distillation column is, ideally, a reversible device [21]. In practice, of course, like all real devices, it is irreversible to a lesser or greater extent. Leaving aside certain irrelevant changes of enthalpy, what happens in a distillation column is that the fluids flowing through it take in heat Q at a higher temperature, T_1, say and reject the same amount at a lower temperature T_2. Now the increase of entropy of the fluids (given that the process is reversible) is

$$\Delta s = \frac{Q}{T_1} - \frac{Q}{T_2} = - \frac{Q(T_1 - T_2)}{T_1 T_2}$$

i.e. the entropy of the fluids is decreased. In fact, the fluids enter as a mixture and leave separated, and their entropy decreases by the entropy of mixing.

Thus the ideal or reversible distillation column is a device in which heat flows reversibly from a higher (T_1) to a lower (T_2) temperature, separating the components of a mixture. The *reversed* reversible distillation column is therefore a device in which heat flows from a lower (T_2') to a higher (T_1') temperature, driven by the reversible mixing of two different fluids; i.e. it is a heat-pumping element and can be used as the basis of a refrigerator.

To make the complete absorption refrigerator, we put a forward and a reversed volumn back to back, the two heat-rejection temperatures being made the same $(T_2 = T_1')$ by running the reversed column at a lower pressure. The fluids A and B separated in the forward column drive by their mixing the heat-pumping reversed column, and a liquid pump is needed to return them afterwards to the high pressure forward column.

This last paragraph gives a highly abstract and fundamental view of an absorption refrigerator, much more powerful as a conceptual tool than the vapour-compression-with-compressor-replaced view. It makes available for improving the performance many general ideas in thermodynamic design and widens our view of what fluids A and B might be like. It increases our insight to a point where difficult questions that would be hard to answer with pencil and paper and plenty of time, can be resolved in the head.

Features of change of viewpoint

It is hoped that readers whose knowledge of thermodynamics is not sufficient for them to understand this illustration fully will nevertheless be able to grasp its general nature. The majority of cases of changed viewpoint contributing to insight suffer from defects as examples; they generally require much study of the problem by the reader before they can be understood, and the change itself is often unimpressive and its value obscure to those not involved. The two chosen illustrations are unrepresentative in that they are short and tidy, and that the new view supersedes the old: more often the new view merely complements the old.

Two features frequently shown by changes of viewpoint are

    (a) a higher level of abstraction, and
    (b) a change of measures, especially to a *task* measure (Section 8.1).

As an example of (b), it is frequently advantageous to regard toothed gearing as a torque-generating device rather than a speed-changing one. Indeed, throughout the design of gearing it is useful to alternate between the kinematic and the statical viewpoint, but this is getting close to the diversification of approach discussed elsewhere (Section 1.5) rather than the development of insight.

Questions

Q.4.1(3). Figure 4.19 shows the principle of a suggested magneto-hydrodynamic

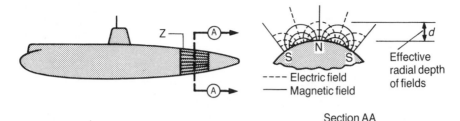

Section AA

Figure 4.19   Magneto-hydrodynamic submarine.

submarine. A portion Z of the hull bears a circumferential pattern of alternating magnetic poles and electrodes. The magnetic poles produce the sort of field shown by the solid lines in the figure; current flows between the electrodes as shown by the dotted lines, the conductor being the surrounding seawater, which is forced backwards, so propelling the submarine forward.

If B is the average magnetic field strength, j the average current density, V the velocity of the submarine and ρ the specific resistance of the seawater, find the condition that the ohmic losses in the seawater are half the useful propulsive work. Assume the current is everywhere substantially at right angles to the magnetic field.

If $B = 1$ Wb/m², $V = 10$ m/s, $\rho = 0.2$ ohm m find roughly the radial distance, d, that the active zone must extend beyond the hull if this condition is to be met; the area of Z is 0.2 times the total surface area, and the drag may be taken as $1.5\,V^2$ N/m² of hull surface.

Q.4.2(1). How must Figure 4.7, illustrating the compressor shaft diameter problem, be modified to cope with cases where the shaft material is of a different density (ρ_1) from the disc material (ρ_2)?

Q.4.3(1). Propose, in algebraic form, figures of merit for materials for the following purposes:

(a) a slender solid strut, to be of least diameter,
(b) a flywheel, to be used as an energy store, of minimum volume,
(c) a fluid for use in a fluid spring of minimum volume,
(d) a fluid for use in a fluid spring of minimum weight,
(e) a material subject to high thermal stress due to fixed temperature differences.

Express your results in a form where high numbers are good, in terms of the following

E: Young's modulus K: bulk modulus
f :   typical allowable stress ρ :   density
$\propto$:   coefficient of thermal expansion

Q.4.4(2). On the same lines as Question 4.3, propose figures of merit in these more difficult cases:

(f) a tape in a gear grinder, which is wound on and off a drum to form a mechanism equivalent to a rack and pinion. The requirement is the minimum stretch in the tape under a given load, the drum size being fixed,
(g) a spin-drier shaft, which must be able to withstand a certain given bending moment, under which its deflection should be as large as possible.

Q.4.5(3). A fluid spring is required of minimum weight and given energy-storage capacity. Its environment is always at the same temperature and operation is very slow, so that the fluid may be regarded as working isothermally. A fluid in a certain thermodynamic state will give the minimum weight of containing pressure vessel. What is that state?

Q.4.6(2). A watch has a balance wheel of linear coefficient of thermal expansion α_W, and the hairspring has a linear coefficient of thermal expansion α_H and Young's modulus E. What must be the rate of increase of E with temperature if temperature changes are to have no effect on time-keeping?

Q.4.7(3). Find the best value of the ratio K of the all-up weight of a man-powered aircraft (i.e. including man) to the weight of the man, on the following assumptions:

    (1) the lift/drag ratio is constant,
    (2) the lift per unit area is proportional to the square of the speed, V,
    (3) the power and weight of the man and the efficiencies of transmission and propulsion are constant,
    (4) the optimum shape of aircraft is independent of K.

Make in turn each of the following assumptions about structural weight:

(a) it is proportional to wing area, and hence to L^2, where L is a typical dimension,
(b) it is proportional to wing area and also to the square root of the load per unit area.

Q.4.8(4). On the same topic as the last question, what would be the optimum value of K on a planet with very dense air, a very high force of gravity, and very powerful inhabitants?

Q.4.9(1). Assuming that traffic in larger towns does not tend to become relatively more localised than in smaller ones, and that densities of population, social habits etc. are also not affected by the size of towns, how will the internal traffic density vary with N, the total urban population?

Q.4.10(2). A coin of radius r is placed on a horizontal surface. A similar coin is placed on top of it, slightly eccentrically so that it overhangs on one side. A third coin is placed on the second, and so on, till the pile contains $n + 1$ coins. What is the maximum amount by which the top coin can overhang the bottom one?

Q.4.11. Sketch a plane mechanism to apply a pure torque to a body. Find a kinematic property which any such mechanism must have and see if this suggests another design.

Q.4.12(2). The inlet to the last stage of a radial outward-flow steam turbine is the circumferential gap between a disc and a ring, both 1200 mm in diameter and spaced apart axially at a distance of 600 mm. The blades are to be supported by the disc at one end and the ring at the other, and the whole rotates at 3000 rev/min.

    Each blade forms a bridge between the ring and the disc, withstanding, not gravity, but centrifugal forces. Find the equivalent length of gravity-resisting bridge and so the form the blade should take.

Answers

A.4.1. $2jp = BV; d = 30$m. A unit cube of water in the active zone has a current j in it, is in a field of strength B, and so has a force jB on it. The *useful* work of propulsion per unit time is thus jBV (see Section 5.2). The ohmic loss per unit volume is ρj^2. If H is the hull area, $0.2\,Hd$ is roughly the active volume, and $0.2\,HdBj$, the propulsive force $= 1.5\,HV^2$.

      Hence $dj = 750$ A/m

Also, since $2j\rho = BV$, $j = 25$ A/m^2, $d = 30$ m. This is clearly impracticable, so it does not matter that $0.2\,Hd =$ active volume has broken down.

A.4.2. Substitute $\rho_1 p/\rho_2 t$ on x axis.

A.4.3. (a) E. (b) f: the kinetic energy stored in unit volume of disc is $\frac{1}{2}\rho V^2$ but the allowable ρV^2 is proportional to f (see Section 3.11). (c) $1/K$. (d) Depends on containment also, but see Q.4.5. (e) $f/E\alpha$.

A.4.4.(f) f: if a material of half the value of E were used, the stresses produced by bending round the drum would be the same if the tape were twice as thick, and hence of the same stiffness in tension. (g) $f^{4/3}/E$: the bending strength requirement means that d, the shaft diameter, is proportional to $f^{-1/3}$. The strain in the shaft under a given deflection is proportional to d, and so the stress is proportional to dE.

A.4.5. Saturated: the figure of merit is $1/K$, which is infinite for a saturated vapour, isothermally compressed. The pressure is constant, and the change in volume can be practically the entire volume of the pressure vessel. The corner work is equal to the real work (Sections 5.8–5.9).

A.4.6. $(2\alpha_W - 3\alpha_H)E$: the fractional increase in the radius of gyration of the balance wheel for unit temperature rise is α_W, so the increase in radius of gyration squared and hence the moment of inertia is $2\alpha_W$ (Binomial theorem). This must be balanced by an equal fractional increase in the torsional stiffness of the hairspring, which increases by $4\alpha_H$ because of the increase in section second moment of area, but decreases by α_H because of the increase in length.

A.4.7. (a) $1\frac{1}{2}$. (b) 2: since the lift/drag ratio is constant, the drag is proportional to K and the propulsive power to KV. In each case KV must be minimised.
(a) Lift $\propto K$, lift $\propto L^2V^2$, structural weight $\propto L^2$, structural weight $\propto K - 1$.

Hence $K \propto L^2V^2$, $L^2 \propto K - 1$,

$$V^2 \propto \frac{K}{K-1}, KV \propto \frac{K^{3/2}}{\sqrt{(K-1)}}$$

Minimise KV and $K = 1\frac{1}{2}$. Similarly for (b).

A.4.8. 1.50–1.67: the wing loading will be high, the structure loading coefficients high, and so the structure will be fully stressed by the loads it carries. In such a case the section areas will be proportional to K, and the structure weight, therefore, to KL.

A.4.9. Proportional to $N^{1/2}$: the area is proportional to N, and hence a typical length of journey to $N^{1/2}$. The product (traffic units $\times$ average distance travelled) is proportional to $N^{3/2}$, and so the amount passing through any unit area is proportional to $N^{1/2}$. This is like the insect breathing problem, but easier.

A.4.10. $r\sum_1^n \frac{1}{n}$:

if you could not do it, try a changed viewpoint, starting from the top.

A.4.11. The equivalent kinematic property is that of a mechanism which can prevent rotation without applying any other constraint. Figure 4.20 shows two possibilities.

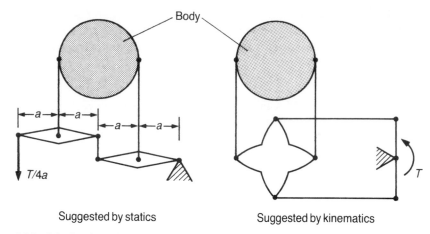

Figure 4.20   Mechanisms for applying a pure torque.

A.4.12. The stresses in any gravity-loaded structure are proportional to $\rho g L$, where ρ is density, g is the acceleratioan due to gravity and L is a typical length. For the blade g is replaced by $\omega^2 r$

$$= (314)^2 \times 0.6 = 59 \times 10^3 \, \text{m} \, \text{s}^{-2}$$

so that the equivalent length of steel bridge, ρ being the same, is

$$0.6 \times \frac{59 \times 10^3}{9.81} = 3600 \, \text{m}$$

or about three times as long as the longest suspension bridge, which is accordingly the only viable form. This blade was Ljungström's experimental suspension blade and its 'sag' took it out to 1600 mm radius, so that it was equivalent to a bridge nearly six miles long. This extreme form gave the large outlet area Ljungström sought as a solution to his matching problem within a single machine (Section 5.3). He based his design partly on the Golden Gate suspension bridge [22].

5 Matching

5.1 Matching: the windlass

As in many other fields of human enterprise, it is helpful when we come upon an aspect of a design problem that has strong affinities with others we have met. Two such aspects, called here 'matching' and 'disposition', will be treated in this chapter and the next.

A rather trivial problem in matching is the design of a windlass for raising buckets of water from a well: Figure 5.1 shows the arrangement, where the drum is such that the rope is wound up with a pitch circle radius r. The drum is turned by hand by means of a handle of throw R. The matching problem lies in choosing r and R so as to enable the human prime-mover to put out the maximum power, let us say, or perhaps to raise the full bucket at a given rate with the least apparent effort—various criteria are possible, and ergonomics is scarcely a fine enough tool yet to distinguish between them. Nevertheless we can determine what is a suitable 'throw' for the winder and put R equal to it; we can likewise specify a suitable value of the force experienced at the handle, F—if it is too high, the winding will be too hard, and if it is too low, winding will be too slow. Then if we put $r = FR/W$, this particularly easy matching problem is solved.

5.2 An extended example of matching: ship propulsion

Suppose that we have to drive a ship at velocity V by means of steam, and we do this

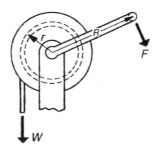

Figure 5.1   Windlass.

by thrusting some of the surrounding water backwards at an absolute velocity v. Then since the necessary thrust T is exerted on water moving backwards at an average velocity $v/2$, the rate of doing work is

$$T(V + \tfrac{1}{2}v)$$

instead of the TV that would be necessary if T were exerted against a fixed object.

The ratio

$$\eta_F = \frac{V}{V + \tfrac{1}{2}v}$$

of these two powers is called the Froude efficiency, and it is clearly desirable that η_F should be as large as reasonably possible. We must keep v small relative to V; if, to fix ideas, we put $V = 10$ m/s and $v = 3$ m/s,

$$\eta_F = \frac{10}{11.5} = 0.87$$

This means we are losing 0.13 of the power used to drive the ship, in the form of kinetic energy of translation in the wake. To reduce this fraction would be desirable, but would generally impose other penalties which would outweigh the gains.

If the steam were expanded through a single reversible nozzle to condenser pressure it might reach a velocity c of, say, 1500 m/s. The ratio $c/v = 500$ gives some indication of the large matching problem that exists in using the small flow of high-energy steam to drive a much greater mass flow of water. A better measure of the water side is provided by the velocity

$$c' = \sqrt{\left\{2\left(V + \frac{v}{2}\right)v\right\}} = 8.3 \text{ m/s}$$

which is the speed the water would have if the specific work done by the propeller were entirely converted into kinetic energy of the water. For the immediate purpose, however, this is not important; all that matters is that the gap to be bridged is a very large one. However, we shall use c' in future instead of v as our typical water speed.

The problem is aggravated by the volume flows involved. The volume flow of sea water is about four hundred times that of the steam in its initial condition, and because of the much lower general levels of velocity in the water the area required to pass it will be much greater than that needed by the steam. It is probable that the radius of the propeller will need to be about 12 times that of the inlet end of the steam turbine (we shall check this later), so that, without gearing, the steam turbine blades would move only one-twelfth as fast as the propeller tips, aggravating the matching problem.

For similar machines we need specific velocities c in the same ratio as the blade speeds. Since c for the steam is 1500 m/s, and the propeller blades move twelve times as fast as the turbine blades, without any form of matching the ideal c' for the water would be $12 \times 1500 = 18000$ m/s. The ratio of this speed to the true c', 8.3 m/s, is

$$\frac{18000}{8.3} = 2170$$

and this is a rough measure of the gap we have to bridge. 2170 is a sort of gearing ratio, in the widest sense, that we have to provide.

With such a large ratio, three methods of matching are needed:

(1) compounding in the turbine,
(2) toothed reduction gearing, and
(3) different blade forms on propeller and turbine.

Crudely speaking, if we have n pressure-compounding stages in the turbine each will have an enthalpy drop and hence a c^2 of only $1/n$ that for a single stage. The value of c is thus reduced by a factor of $1/\sqrt{(n)}$, and so with 20 stages we should achieve a ratio from this source of about 4.5, leaving

$$\frac{2170}{4.5} \simeq 480 \text{ to go}$$

Of course, regardless of the matching problem, there are imperative reasons why the turbine should be compounded, both fluid-dynamical and mechanical, so this much matching costs us nothing. In the early days of turbine ships, however, when attempts were made to avoid toothed gearing, very large numbers of stages were used [23].

A contribution from (3) can be had by making the turbine blades of such a form that they extract the maximum possible energy from the steam relative to their velocity u, i.e. such that the ratio u/c is as low as possible, consistent with efficiency. The form of the propeller blades on the other hand should be chosen so they do little work relative to their velocity u', i.e. so that u'/c' is high. In Figure 5.2, all forms of turbomachinery are shown plotted against two parameters, one of which is u/c. The lowest value of u/c attainable with efficiency (excluding velocity compounding or a stage followed by a diffuser) is a little under 0.5, given by the impulse turbine.

With propellers very high values of u/c can be obtained, but for reasons which will not be discussed here the practical limit for the ship is about 4. Blade form differences, obvious in the contrast between the strong curvature of the turbine blade and the flattened form of the propeller blade, can thus give us a ratio of about 8, leaving a quotient of about 480/8, or about 60 to be taken care of by the toothed gearing.

It must be understood, of course, that this broad and simple study of the ship-propulsion problem is not of any practical use. But it is hoped that it will be of substantial value as an example of the sort of abstract and yet highly pertinent treatment of a matching problem which generates insight and leads to a rapid and quantified grasp of less familiar or less thoroughly studied design situations (see Question 5.2).

The parametric plot (Section 2.4) in Figure 5.2 has much in common with the insets of Figures 2.4–2.8 and may also be regarded as a graphical piece of the engineering repertoire (Section 2.3). The other axis is of

$$\frac{4Q}{\pi D^2 c} = \psi \text{ (say)} \tag{5.1}$$

where Q is the volume flow, D the diameter, and c has already been explained. This is a measure of the *size of the volume flow* relative to the machine that handles it. To reduce as much as possible the mismatch owing to the diameter of the propeller being much greater than that of the turbine, we must make the propeller as small as possible (i.e., with a high ψ) and the turbine as big as possible (with a low ψ). It can easily be estimated that the propeller will have a ψ' of about 1.2, and we can put, using primes to denote propeller values,

$$\frac{\psi}{\psi'} = \frac{Q}{Q'} \cdot \frac{D'^2}{D^2} \cdot \frac{c'}{c}$$

or $\dfrac{\psi}{1.2} = \dfrac{1}{400} \times 144 \times \dfrac{8.3}{330}$ (say)

(allowing for the reduction in c by compounding)

or $\psi = 0.011$ (say)

Looking at Figure 5.2, it will be seen that this is possible in combination with $u/c = 0.5$, though nearing the limit where efficiency falls rapidly due to the over-short blades, windage etc. (Notice that this limit can be exceeded where there is a free surface, as in a Pelton wheel, and the windage is due only to a fluid of much lower density than the working fluid; then high losses are not associated with partial admission.)

Notice that the solution of this ship-propulsion problem involves taking the machines from the (practicable) bottom-left-hand and top-right-hand corners, i.e. from the extremities of the field, and pushing up against a number of limits such as turbine first-stage efficiency, propeller efficiency and cavitation. This is a very common feature of matching as of other design problems—a judicious pressing towards the limits in all helpful directions.

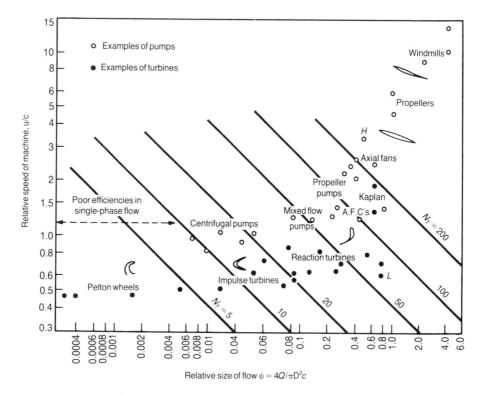

Figure 5.2   Family of turbomachines. Profiles of blades of typical form are shown: towards the top right they become thinner and flatter, reflecting the lower specific work relative to the blade velocity. H = hovering helicopter, L = turbine stage of Q. 4.12

5.3 Matching within a single machine

One of the aggravating circumstances in the ship-propulsion problem was the large ratio of the water volume flow to the steam volume flow. In the turbines of the turbo-alternators of power stations the same problem arises owing to the great increase in specific volume between the inlet and outlet steam conditions, which may be about 1600 times. The situation is eased by the short circuiting of the last stages by steam withdrawn for feed-heating purposes, so that the mass flow at inlet is substantially higher than that at outlet, giving a volume ratio of, say, 1100 to cope with.

If we put $Q = \pi D^2 c\psi/4$ (from equation 5.1) and use the suffixes I and O to denote inlet and outlet respectively,

$$\frac{Q_O}{Q_I} = \frac{\psi_O}{\psi_I} \cdot \frac{D_O^2 c_O}{D_I^2 c_I} \tag{5.2}$$

It is desirable that this ratio should be large. If the angular velocity ω is the same for all stages, i.e. we have only one shaft, then $D = 2u/\omega$, and equation 5.2 can be written

$$\frac{Q_O}{Q_I} = \frac{\psi_O}{\psi_I} \frac{(u_O^2/c_O^2)c_O^3}{(u_I^2/c_I^2)c_I^3} = 1100 \tag{5.3}$$

Thus again we need to go from a low ψ and a low u/c at inlet towards a high ψ and a high u/c at outlet, sloping upwards to the right in Figure 5.2. We might scrape 40 from the ratio of ψs, perhaps another 3 from the (u/c)s and a little from the cs. We shall still be short of our target by the best part of a power of ten.

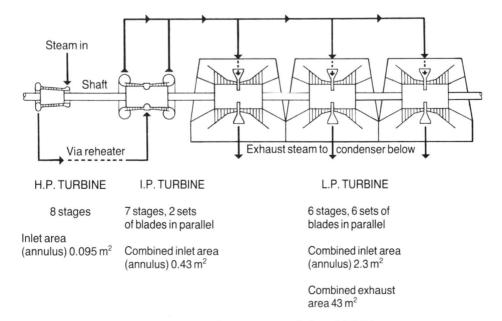

H.P. TURBINE	I.P. TURBINE	L.P. TURBINE
8 stages	7 stages, 2 sets of blades in parallel	6 stages, 6 sets of blades in parallel
Inlet area (annulus) 0.095 m²	Combined inlet area (annulus) 0.43 m²	Combined inlet area (annulus) 2.3 m²
		Combined exhaust area 43 m²

Figure 5.3   Multiple L.P. turbine. (By kind permission of GEC.)

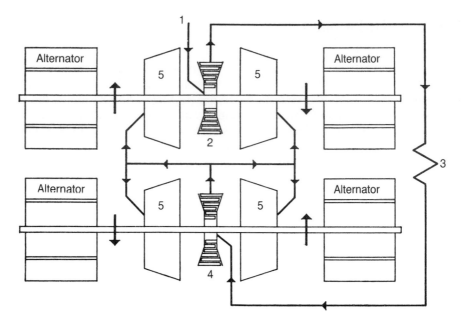

Figure 5.4   Stal-Laval Quad Turbo-alternators. High-pressure steam enters at 1 and flows outwards through the contra-rotating blading of the radial-flow H.P. turbine 2. It continues via the reheater 3 to the intermediate pressure turbine 4, which is also of the contra-rotating radial-flow type. The flow is then divided into four and enters the four axial-flow low-pressure turbines 5. Since the *effective* speed of the contra-rotating H.P. and I.P. turbines is twice that of the L.P. turbines the assistance with matching given by this arrangement is equal to that given by dividing the flow into $2^2 \times 4 = 16$, against the 6 of Figure 5.3.

The most hopeful ways of bridging the remaining gap are

    (1) to have more than one ω, i.e. to have more than one shaft,
    (2) to have several turbines in parallel in the later stages.

Figures 5.3 and 5.4 show two solutions to the matching problem between first and last stages in a turbo-alternator turbine.

5.4   Further aspects of ship propulsion

The problem of matching within the turbine arises also in the steamship, and is generally solved by using two shafts with different angular velocities. Since reduction gearing is needed in any case, there is no cost penalty involved; indeed, there is a gain (Section 8.2), and the shorter, broader plan obtained by cutting the turbine into two lengths and putting them side by side suits the available space much better.

Before about 1958, the fuel costs of steam turbines and i.c. engines for ship propulsion were about the same, the steamship's boilers burning more but cheaper oil. When it became possible to use boiler fuel in the big slow-running marine i.c. engines, steam was at a critical economic disadvantage. For the largest powers, however, required in the supertankers, advanced steam plants have been designed with fuel consumptions which have been pared down to no more than about 15 per cent higher than the engines. The question has been studied extensively of how to reduce or eliminate this remaining gap.

It is not practicable, by known means, to reduce much further the fuel consumption of the steam plant per shaft kilowatt hour, but if the propulsive efficiency of the propeller could be increased, the same SHP would go further. Now the big marine i.c. engines run at 100–110 rev/min, and this gives too high a value of u/c for propeller efficiency, which would be substantially better at 75–80 rev/min. The i.c. engines could be made to run at this speed, but since their power is roughly proportional to the product of their swept volume and their speed, they would have to be 33 per cent bigger. For the steamship to reduce its propeller speed, however, only requires a somewhat costlier reduction gear (Section 8.2). This is an example of a general principle we shall call 'exploitation of assets' (Section 8.1); having been forced to adopt this expensive complication, the reduction gear, the designer should not stop short at providing the propeller speed of 100–110 rev/min traditional for the i.c. engine, but press on down, as it were, to the optimum speed, where the rival system is only able to follow at much greater expense. Having accepted gearing, it should be exploited to the full and not simply used to overcome the original difficulty.

However, it is unlikely that the story will end here; already good results are being obtained with boiler fuels in i.c. engines of medium speed (in this context, say, 350 rev/min). Such engines must be geared, and so are in the same position as steam turbines as regards using either lower-speed propellers, or another alternative giving similar improvements in efficiency, contra-rotating propellers (Section 8.3). They retain the basic advantage over steam of lower fuel consumption per SHP hour; they have advantages over the big 'cathedral' engines of smaller cost, weight and size, and easier repairs, provision of spares and so forth. If certain problems are overcome, this third contestant may triumph over the other two.

5.5   Specific speeds: degrees of freedom

Some readers who know something of specific speed may be troubled by Figure 5.2, which might be called a Procrustean chart because of the ruthless way in which all turbomachines have been made to fit it. Figure 5.2 is plotted against two parameters; how is it that for water turbines only one parameter, specific speed, will suffice?

For any given requirement for a machine, we could choose u/c and ψ anywhere in the workable region of Figure 5.2: we have two degrees of freedom of choice. In the steamship problem, the requirement of matching turbine and propeller with as little gearing as possible meant choosing them as far apart as possible in value of

$$\psi\left(\frac{u}{c}\right)^2 \text{(see equation 5.3)}$$

which is simply shown to be proportional to the square of the specific speed. Now we have still applied only one condition (namely, a difference of specific speed), but because of the location of the boundaries of the workable region this has dictated our solution completely; in fact, we are reduced to the two extremities of the region perpendicular to the lines of constant specific speed, and we still get only a factor of 8 or so towards our matching requirement of about 2000.

Where there is no dominant matching requirement as in a water turbine for electricity generation, we have two degrees of freedom of choice (see Section 9.2). These must be exercised to reduce the capital cost of the machine and increase its efficiency as much as possible. The general effect of this is to concentrate the economical solutions with no matching requirement along a line, and the position of a point on this line requires only one co-ordinate to specify it, specific speed or ψ, say (u/c would not do because for part of the range the line has constant u/c).

To see a little of the reasons for this concentration of the unmatched solutions in a line (or, rather, a narrow belt), consider a few examples. A medium-head, high-flow turbine could be built as a Pelton wheel, with a value of ψ perhaps 0.01 times that of the reaction turbine which would be the usual choice, and so, since Q and c remain unaltered, 10 times the diameter. Since the reaction turbine might be, say, 8 m in diameter, it is not difficult to see that the capital cost of the Pelton wheel would be excessive. If, on the other hand, we try to invade the home territory of the Pelton wheel with the reaction turbine, the mechanical problems imposed by the high pressure, particularly in the variable nozzles, become incompatible with an efficient solution.

In the range in which the Pelton wheel and the high-head Francis turbine are in competition, and the former suffers from its large size, it should be noted that the $1\frac{1}{2}$ power law holds (see Section 4.5), so that if we use three smaller Pelton wheels, each will be of $(1/3)^{1/2} = 0.577$ the diameter and the total weight approximately

$$3 \times \left(\frac{1}{3}\right)^{3/2} = 0.577$$

and this big reduction will be strongly reflected in the capital cost.

In principle it is wrong to optimise the turbine apart from the alternator but the effect of the latter is generally small, for reasons given in Section 4.7. The interaction of the turbine speed and alternator cost is not powerful (though sometimes influence arises from questions of attitude; i.e. whether axes are horizontal or vertical).

Many important questions have been begged here, the effects of Mach number and Reynolds number, cavitation, control aspects and so on. Above all perhaps it should be noted that propellers, ducted fans, windmills and some bulb turbines (Röhrturbinen) have in effect an extra degree of freedom since the ratio of the specific work to the kinetic energy or apparent kinetic energy can also be varied. On the other hand, the optimum hub:tip ratio is usually the smallest that can be achieved, so that one of the other degrees of freedom is virtually eliminated, and a two degree of freedom plot is not too 'Procrustean' in its effect.

This discussion of turbomachinery has illustrated, not only matching, but important ideas like exploitation of assets, recognition of limits, division of function among a number of similar units, dimensional methods, and, above all, the generalised way in which it is possible to view this large and very important section of the engineering repertoire.

5.6   Matching of a spring to its task

The windlass and ship-propulsion are examples of matching by gearing in a broad sense, where the application of a constant ratio to an angular velocity or a force is all that is needed. The essential element of gearing in a narrow sense is the lever or the pair of toothed wheels, sprockets and chain, pulleys and belts, friction discs etc., which are all embodiments of the lever principle, made continuous in operation.

In this section an example of gearing in a broad sense will be studied, the selection of the form of a spring which is to exert a maximum force F under a compression in length of a. We saw in equation 4.9 that the energy stored per unit volume of the most highly stressed material in a helical spring, working at point C in Figure 3.10, was

$$\frac{(1+v)f^2}{4E}$$

and that the efficiency of use η_u was $\frac{1}{2}$.

Thus, ignoring dead turns and stress concentration, the energy that can be stored is

$$\frac{(1+v)f^2}{8E}v$$

where v is the volume of material in the spring. This must be at least equal to the energy required to be stored in the spring, which is $\frac{1}{2}Fa$. We thus know that the minimum volume of stuff needed is

$$v = \frac{4FaE}{(1+v)f^2} \tag{5.4}$$

If we cut off a length L of wire of any radius r within a wide range such that the volume has this value, then it can be made into a spring having the required characteristics. If the coils are flattish, as in a practical spring, the wire is nearly in pure torsion, with a constant torque FR in it and a constant twist per unit length, λ, say. The reduction in length of the spring due to the twist in a short length ds of the wire is

$R\lambda ds$ (see Figure 5.5)

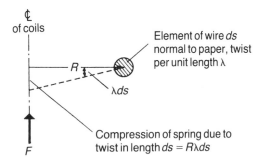

Figure 5.5   Wire of compression spring.

so that the total compression is

$$R\lambda L = a \tag{5.5}$$

Now the twist per unit length is given by

$$\lambda = \frac{\tau}{Gr} = \frac{2\tau(1+v)}{Er} \tag{5.6}$$

where G is the shear modulus, τ is the shear stress in the surface, and v and E have their usual significance in this context (see Section 4.5).

If the material is working at point $C(-f/2, f/2, 0)$, $\tau = f/2$ so that from equations 5.5 and 5.6,

$$R\lambda L = R\frac{f(1+v)}{Er}L = a$$

or $\quad R = \dfrac{Era}{f(1+v)L} \tag{5.7}$

Since L is given when r is known and v fixed by equation 5.4, by the relation

$$v = \pi r^2 L \tag{5.8}$$

equation 5.7 gives a value of R corresponding to each r which suitably 'gears' the wire material to its task. If the wire is very thin, then R will be very small and a design impossible; if the wire is too thick, then R will be too large relative to L to make a satisfactory spring.

Notice that we could have derived an alternative relationship for R by considering the torque in the wire, FR, and relating it to the stress τ by

$$FR = \frac{\tau J}{r}$$

where J is the polar second moment of area of the wire cross-section, which would lead to the same result. In just the same way, if we have a 1000 kW machine driven by a 1000 kW prime mover, we can choose the gearing between to equate either torques or speeds, and then the remaining quantities, speeds or torques, will automatically be equal. The energy storage capacity of the spring fixes the quantity of material, and then if either the force or the deflection is matched, the other requirement is automatically fulfilled.

As an alternative, we could use any spring of the right energy storage capacity and a lever to match it, as in Figure 5.6(a). Suppose P is the full load and d the full-load deflection of the spring to be 'geared' by means of the lever, then $Pd = Fa$ by the equal energy storage requirement, and from the figure, $Pb = Fc$, so that $d/b = a/c$, a result that could have been found equally well by considering the travels a and d.

By using a non-linear lever, such as the cranked one of Figure 5.6(b), we can make a linear spring of constant stiffness into a non-linear one: in Figure 5.6(b), as the deflection increases, the arm c decreases, the 'gearing' from F to the spring increases and so the stiffness at F grows larger. This is useful in such applications as car springs, where it is desirable that the stiffness of the suspension should increase with load in order to keep the natural frequency of vibration about constant.

A spring of this sort, with a constant ratio of stiffness to load, will have an exponential variation of force with deflection, as in Figure 5.7. This is the first

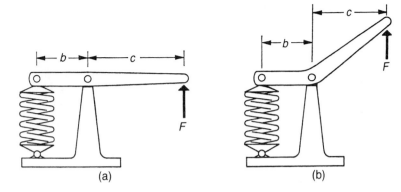

Figure 5.6   Spring 'geared' by lever.
(a) point matching, (b) curve matching.

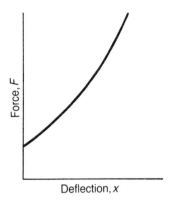

Figure 5.7   Exponential characteristic of suspension spring.

example we have used yet of matching a *shape*, rather than a point value. The next
section deals with a similar problem.

5.7   Matching in thermodynamic processes

If, in a heat exchanger, a fluid A is giving up heat to a fluid B, then the temperature
T_A of A must be greater than the temperature T_B of B at every point. If the
difference is too small, the heat exchanger will be too big and expensive, but if it is
too large, the large irreversibility will be apparent in reduced efficiency. This
subject was dealt with in Section 3.5 for fluids of substantially constant specific heat
at constant pressure (c_p), but matching problems arise when the specific heat of one
or both fluids varies extremely, as when a pure substance is being heated to its
boiling point, evaporated and then superheated in a boiler.

Figure 5.8 shows heating and cooling curves for the heat exchange in the boiler of
an early gas-cooled reactor (GCR) nuclear power station. Along the x axis is
plotted specific enthalpy, and along the y axis absolute temperature to a non-linear
scale which is explained later. Curve A shows the variation of enthalpy with

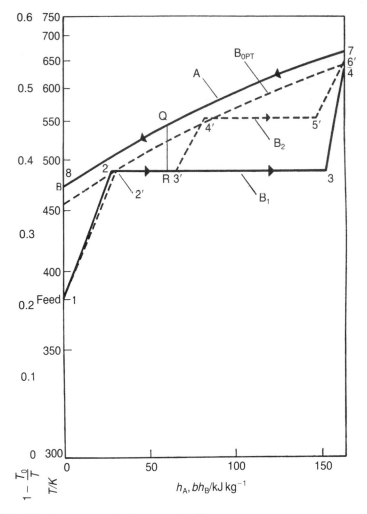

Figure 5.8   Heating curves, dual-pressure boiler.

temperature of the carbon dioxide which is evaporating the water, cooling itself in the process (indicated by the arrow head from right to left). Curve B_1 (solid line) shows the variation of enthalpy with temperature in a first design in which steam is raised at a pressure of 2.1 N mm^{-2}; from 1 to 2 water is being heated at a constant pressure of 2.1 N mm^{-2}, from 2 to 3 it is being evaporated at the same pressure and constant temperature, and from 3 to 4 it is being superheated.

   Since the mass flow of water and that of carbon dioxide are not the same, the enthalpies shown cannot be specific for both flows. For A the specific enthalpy h_A of the carbon dioxide is used, referred to a zero at the outlet conditions, while for B_1 bh_B is plotted, where b is the ratio of the mass flow of water to that of the carbon dioxide, and h_B is the specific enthalpy of the water. The zero for h_B is taken at the inlet conditions. Pressure drops and heat exchange with the surroundings are ignored.

   By the steady flow energy equation (first law), then decrease in h_A from 7 to 8 (see the figure) is equal to the increase in bh_B from 1 to 4, and since we have put 8 directly over 1 in the figure, then 7 is directly over 4. Any two adjacent points Q, R, one on the carbon dioxide and one on the steam side, lie directly over one another in the figure, by applying the same argument to the part of the heat exchanger, from QR to either end. Then the condition that T_A is always greater than T_B is that curve A always lies wholly above curve B_1.

Now the scale of T has been chosen so that the total area under the A curve, right down to the heat rejection temperature T_0, gives the available energy lost by one kilogram of carbon dioxide, and the area under B_1 represents the available energy gained by the corresponding b kg of steam (see Section 4.5). The area between A and B_1 is the lost work per kg of carbon dioxide resulting from the temperature difference in the heat exchanger, and owing to the poor matching of the curves A and B_1 it is very large, about 0.20 of the area under A. Another way of putting this is

$$\text{boiler thermodynamic efficiency } \eta_1 = \frac{\text{increase of available energy of water}}{\substack{\text{decrease of available energy of} \\ \text{carbon dioxide}}}$$

$$= \frac{1 - 0.20}{1} = 0.80$$

Curve B_2 is the heating curve for two separate flows of water, a small one at 2.1 N mm^{-2} (LP) and a larger one at 6.5 N mm^{-2} (HP). The parts of B_2 are

1'–2' HP and LP water being heated
2'–3' LP water evaporating to steam
       HP unchanged in temperature
3'–4' HP water being heated
       LP steam being superheated
4'–5' HP water evaporating to steam
       LP unchanged
5'–6' HP and LP steam being superheated.

By using two separate pressures and so producing a heating curve with two steps which fits A much better, the lost work, the area between A and B_2, is only 0.12 of the area under A, and the new boiler efficiency is

$$\eta_2 = \frac{1 - 0.12}{1} = 0.88$$

a substantial improvement.

Even if we could make the B curve exactly the same shape as the A curve, we would still have to keep a temperature difference of economic size, as in B_{OPT} in Figure 5.8. This might account for about 0.04 of the area under A, so that a fairer assessment of the matching losses would be

	Matching loss as fraction of area under curve A
B_1	0.20–0.04 = 0.16
B_2	0.12–0.04 = 0.08

This illustrates, with fortuitous exactness, a general principle of matching: that, if we adopt staging or cascading to reduce this kind of matching loss, the remaining loss will be roughly equal to the original loss divided by the number of stages, N. Thus going to the dual pressure from the single pressure steam system we reduce the matching loss from 16 per cent to 8 per cent, and if we went to three pressures we might expect a reduction to 5.3 per cent, a further gain of only 2.7 per cent. This same behaviour is exhibited by the joint efficiency of fir-tree roots and some riveted joints, and by other examples of thermodynamic matching such as regenerative feed heating, reheating in turbines and pressure staging in refrigerators.

With increased reactor temperatures the advantage of dual pressure cycles has disappeared. The whole of curve A is shifted much higher.

The importance of matching in thermodynamic design is paramount. Whether we wish to produce electricity via steam, to liquefy natural gas for transport by sea, to refine crude oil, to effect air conditioning by means of an absorption refrigerator (see Section 4.10) or to desalinate seawater, the processes we first conceive can be systematically improved in performance by looking for imperfections of matching and reducing them. To heat water from condenser temperature with high-temperature furnace gases is uneconomical because of the large temperature mismatch—we use steam tapped from the turbine instead, at several intermediate stages so as to match the smooth heating curve of the water with a series of steps in the combined cooling curve of condensing steam. The throttling of a liquid refrigerant in stages, the vapour at each stage being returned to an intermediate inlet of the compressor, is the functional reverse of this, with the steps in the heating instead of the cooling curve.

More strictly speaking, we improve thermodynamic processes in efficiency by the systematic removal of irreversibilities, and irreversibilities are due to either friction or mismatching, usually of temperature but sometimes of pressure or composition of mixing fluids. Since the frictional irreversibilities are not usually tractable we are left with mismatching. To see why the throttling of a saturated liquid is irreversible mainly due to temperature mismatch, consider the equivalent device shown in Figure 5.9(b) in which the liquid is cooled *before* throttling to the saturated temperature $T_{sat\,2}$ obtaining *after* the throttle, the cooling being effected by the products of throttling: it will be found that most of the irreversibility in this completely equivalent device is due to temperature mismatch in the heat exchanger.

Figure 5.10 shows the cooling curve of natural gas being liquefied in a very advanced piece of matching. The refrigerant is a mixture, which accounts for the curious form of its heating curve. It is interesting to note that the matching in this process involves composition as well as temperature, and that the 'equivalent device' of Figure 5.9 is no longer equivalent to the throttle but much superior to it

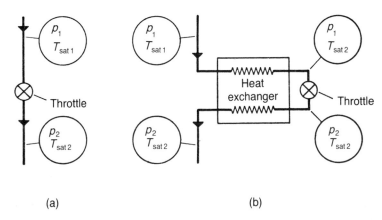

(a)   (b)

Figure 5.9   Equivalent of isenthalpic throttle. (a) Isenthalpic throttle, (b) Equivalent for pure substances.

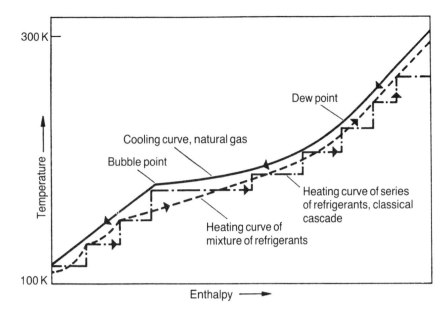

Figure 5.10   Liquefaction of natural gas against a mixture of refrigerants, showing the superiority of matching over that of a more complicated cascade of pure refrigerants. (By kind permission of Technip, Paris.)

when mixtures of substances are being handled, so much so that designs not using it would not be viable.

Matching in all fields should be taken only so far as is economically justified. This will be much further in refrigeration than in prime movers because as a rule only fuel and boiler costs are being saved in the latter, but in the former it may well be fuel, boiler, turbine, compressor, and heat-exchanger costs. Also the shape of the temperature function $1 - (T_0/T)$ in Figure 5.8 means that temperature mismatches of a given number of degrees are more important at lower temperatures.

5.8   Two old cases of matching

(1)   Bows

There is some suggestion, from the Mongol victories for example, that the reflex bow of the ancients and the East was more powerful than the longbow [24]. The conspicuous feature of the reflex bow is its highly prestressed state when strung (Figure 5.11(a) and reference 25). Figure 5.11(b) shows two hypothetical curves of force (F) against draw (x) of bows. Each has the same force (or 'weight') when fully drawn, and the same draw, so they are in effect matched to the same archer (having a fixed 'corner work'—cf. corner power, Section 4.6). The heavily prestressed reflex bow will have a sharper initial rise of F and a smaller stiffness later, giving a curve more like R than L, and with more area under it. As the area under the curve is the stored energy, the 'power' of the reflex bow should be greater for the same weight and draw.

It may be objected that the heavy prestressing of the reflex bow means that only a

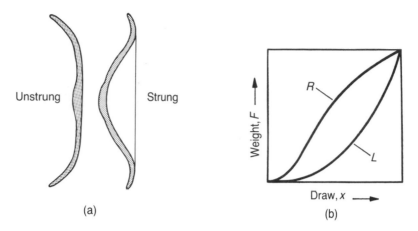

Figure 5.11   Bows. (a) Reflex bow, (b) Characteristics (hypothetical).

fraction of the strain energy of the bow material is available to drive the arrow, and this is true. As the spring material used had a very high strain energy capacity, however, the bow was not heavy: the construction was laminated, with sinew on the tension side, wood in the middle and horn on the compression side—an excellent though primitive example of composite construction to secure the best available material properties at every point for the condition obtaining there [24]. The best animal protein fibres will store about 17 times as much energy per unit mass as spring steel.

    An interesting point about all bows is that the ratio of the velocity of the arrow to the velocity of the tips of the bow increases rapidly as the string nears the bow. Kinetic energy which is stored in the bow material immediately after release is mostly imparted to the arrow via the string, so that very little kinetic energy is left in the bow afterwards.

(2)   Reciprocating direct steam-driven water pumps

Early in the 20th century, direct-acting steam-driven water pumps were used, of the sort shown diagrammatically in Figure 5.12. The force exerted on the piston assembly by the steam is uniform up to the cut-off C, and then decreases as the steam expands (Figure 5.12b). The resisting force of the water will be constant, provided the back pressure on the pump is uniform. If the average steam force is equal roughly to the water force, as must clearly be the case, there will be an excess of steam force at the beginning of the stroke and a deficiency at the end. In a high-speed machine with a crankshaft, this mismatch would be smoothed out by a slight fluctuation in the kinetic energy of the flywheel. In the slow-speed pumps in question, however, it was desirable to match this unbalanced force by the two air cylinders shown in the figure, which acted simply as springs. The three forces on the piston assembly, all plotted as positive when acting to the right, can be seen to nearly cancel one another throughout the stroke, giving the required matching between the steam and water forces.

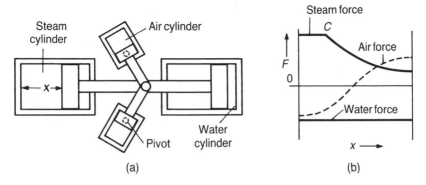

Figure 5.12   Direct-acting steam-driven water-pump.

5.9   Car handbrake

Consider the problem of incorporating a handbrake in a motor-car disc brake by means of a manually operated mechanism acting on the same brake pads as the footbrake. This, like the design of a bow, is a corner-work problem.

Figure 5.13(a) shows the principal parts concerned, the disc itself and the two opposed pads of friction material which are clamped onto the disc by nipping forces, F, when the handbrake is applied. When the footbrake is applied the pads are also clamped, but we can largely ignore this aspect for the purposes of this example, though not, of course, in the design.

It is required that the force, F, shall have a certain value (F_N) when the handbrake is fully applied in order that it shall be capable of holding the car on some specified gradient, say, 1 in 4. It is also necessary that the clamping mechanism shall have sufficient travel d to take up clearance between the pads and the disc and elastic deflection under F_N in the caliper or other structure closing the force path (shown dotted in Figure 5.13(a)). Note: the drawing of such force paths is often helpful in visualisation [44]. While the elastic deflection under F_N is constant, the clearance or backlash between the pads and the disc is variable, so that upon applying the brake the force/travel curve may take either of the forms 1 and 2 in Figure 5.13(b), or any

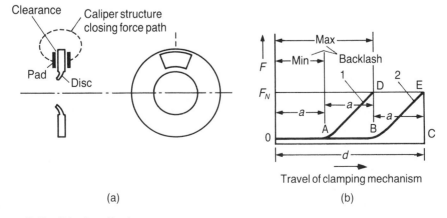

Figure 5.13   Disc handbrake.

similar shape in between. The flat part at the beginning of each curve is the portion of the travel in which clearances are taken up, and after that is a triangular piece in which further travel is solely due to the elastic deflection of the structural parts.

To fix ideas, let us assume that the clearance between the pads and the disc before applying the handbrake can range from $a/2$ to a on each side, and there is no other backlash in the system. Let us further assume that there is no friction or elasticity in the system except the elasticity in the caliper, which deflects a under the load F_N. Then the actual work done by the driver is the same as that done at the clamping mechanism, and Figure 5.13(b) has the proportions shown, where OA = AB = BC = a. Whether for curve 1 or curve 2, the actual work done by the driver per brake is $\frac{1}{2}aF_N$.

Suppose there is a fixed mechanical advantage R between the handbrake lever and the clamping mechanism, so that the force per brake P exerted by the driver must be F/R at all times, with a maximum value $P_N = F_N/R$, then the travel of the handbrake lever in the worst case (curve 2) is $3aR$ and the corner work (Section 5.8) demanded of the driver, defined by (maximum force × travel), is

$$3aR \times \frac{F_N}{R} = 3aF_N \text{ per brake}$$

This is so large in practice as to be an embarrassment in a real system with friction, elasticity and backlash in the other parts. Observe that corner work, like real work, is not changed by gearing of a fixed ratio R, but unlike real work, it is changed by a variable R.

An ideal mechanism is theoretically possible in which the handbrake lever would always travel its full distance at a uniform value of P. Here the force/travel relationship at the driver's end is rectangular and the corner work is therefore equal to the actual work or $\frac{1}{2}aF_N$, a reduction to one-sixth which would more than overcome the problem of 'heavy' handbrakes. This ideal mechanism would be perfectly matched to the load, whether it followed curve 1 or curve 2 of Figure 5.13(b) or any intermediate shape.

Suppose the pad clearance to be at its minimum, so that the resistance at the brake follows curve 1 in Figure 5.13(b). Then while the clamping mechanism travels from O to A, R will be very low in the ideal device, and the hand lever will move only a small distance. As resistance is encountered at A, R will grow proportionally to F to give a constant value of P. If the clearance is at its maximum, so that curve 2 is followed, this growth of R will not begin till point B is reached.

Apart from the possibility of finding an ingenious 'cut price' embodiment (see Section 2.12) of this ideal solution, it must be regarded as an impractical (though not impracticable) ideal. It does give us the absolute limit of what is possible, however; since a reduction to one-sixth is theoretically possible, it is not too optimistic to hope that a reduction to one-half is practically possible. Let us look for easier ways of making more modest improvements.

Suppose instead of the continuously variable R required by the ideal mechanism, we contented ourselves with two values of R, R_1 and R_2. Then the clearance would be taken up at the small value R_1, for very little movement of the hand lever. When the resistance increased, the mechanism would 'change gear' to a high value R_2 at which the remainder of the action would be completed. If we ignore the little bit at R_1 the travel of the hand lever would be aR_2 and the maximum value of P, F_N/R_2. Thus the 'corner work' would be aF_N, an improvement of three times over the fixed R mechanism.

This is still a rather advanced mechanism. The ideal version matches itself perfectly to whatever force/travel relationship arises, curves 1, 2, or in between. This second version matches itself within the limits of its two 'gears', R_1 and R_2; it still requires some device which reacts to a certain level of resistance by changing gear. Figure 5.14 shows a hypothetical form of such a mechanism.

A smooth bore in a block G houses a sliding plug H and block L, the latter acting directly on a pad and thus being subjected to the resistance F. The plug H is hollowed out to a thin-walled cylinder at the end adjacent to L and a rubber cylinder K is bonded in: the other end contains a small sliding piston J, which is pushed by the hand lever via a mechanism of fixed mechanical advantage R_1.

When F is small or zero H and L slide freely in the bore and $R_1 P = F$. As F, and hence P, increases, however, the squashed rubber expands the skirt of H, which locks in the bore. The rubber then behaves rather as the fluid in a hydraulic ram, with J the loading piston and L the ram, so that the force $R_1 P$ is magnified roughly in the ratio of the cross-sectional area of J to that of K.

This 'cut-price' solution is still rather elaborate, however. Let us remove altogether the need for behaviour sensitive to resistance by using a mechanism in which R varies in a fixed way with travel, matched to curve 1 as far as point B and keeping R unchanged thereafter, i.e. a mechanism perfectly matched to the imaginary load curve OADE, which is the envelope of all possible load curves. It is clear that this arrangement provides more mechanical advantage than is necessary for all load curves except 1.

Since this last mechanism is perfectly matched to the load curve OADE, its corner work is equal to the actual work of OADE, or $\frac{3}{2}aF_N$. Thus a very useful reduction to one-half has been achieved with a relatively simple approach. This shape OADE is not easy to match, and one existing design which uses this kind of matching has a cranked lever, similar to that of Figure 5.6(b) in principle. However, the problem has been over-simplified in this account and a smoother matching curve may well be desirable when other effects are taken into account. The cranked lever in question is also a good example of exploitation of assets (see Section 8.1), since it serves another purpose.

We have tried three levels of matching to reduce the 'corner work' demand of the handbrake—the best possible matching, a simpler version which still adapts itself to the actual variations of load curve, and the simplest, which takes no account of variations in load curve; the reductions effected are to one-sixth, one-third, and one-half respectively. The first two approaches might be called 'adaptive' and are rather elegant but difficult to embody in a practical form.

Before leaving this example, another way of improving the handbrake matching

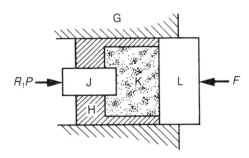

Figure 5.14   Mechanism giving ideal handbrake characterics.

should be considered: we might, as in the steam-driven pump of Section 5.8, store energy in the first part of the hand-lever travel to be released in a way helping brake application in the second part. If we used a spring, we would merely be adding to the force/travel curves of Figure 5.13(b) a fixed curve of zero area, like the air-force curve of Figure 5.12.

On the other hand, the energy might be stored in a flywheel, which behaves rather like an 'adaptive' mechanism in that it generates torque to overcome any resistance so long as it continues to have kinetic energy left in it. The flywheel may be regarded, in fact, as a simple and perfect adaptive device in this sort of context; while unsuitable for several reasons for the brake, it is of course used in the same matching role on the starter of a car, the starter motor armature providing the flywheel.

We have emerged from this study of the matching problem of a handbrake without any very good solutions, the best being the fixed variation of R with travel. For further improvements attacks might be made on the caliper flexibility and also, an aspect which has been left out but is in practice most important, the flexibility of the transmission between hand lever and brake (see Question 5.11).

5.10   General remarks on matching problems

Many types of matching problem have been omitted for the sake of brevity and others, no doubt, because the writer has not met them or has forgotten to consider them. One important category is off-design-point matching, which is especially important in turbomachinery (see Question 5.8), and another is the special problems of traction. Attempts have been made in the questions that follow this chapter to fill a few of the more conspicuous gaps and to show a little more of the range of guises in which a need for matching may present itself and the means that can be adopted to meet it.

Questions

Q.5.1(1). An electrical power source may be regarded as a generator of voltage, V, in series with a resistance R_1. This source is to be used to power a device that behaves as a resistance R: what value must R have so that the power expended in the device is a maximum?

Q.5.2(2). A cryogenic plant is to have a turbine that expands a flow of helium from 40K and 0.38 N mm^{-2} absolute to 26K and 0.11 N mm^{-2} absolute. It is proposed to use the power from this turbine to compress the same flow of helium from 300K and 0.11 N mm^{-2} absolute to as high a pressure as possible, by means of a centrifugal compressor. Can the turbine and compressor be matched without gearing?

Q.5.3(3). Figure 5.15 shows a simple letter balance, in which the only weights that need be considered are those of the letter, W and the balance weight, w. The extremes to be weighed are 1 and 16 units, and the relative sensitivity (rate of change of θ with fractional change in W, i.e. $W(d\theta/dW)$) is to be the same at each of these values of W. Find values of w and ϕ that fulfil these conditions and make the minimum sensitivity as high as possible.

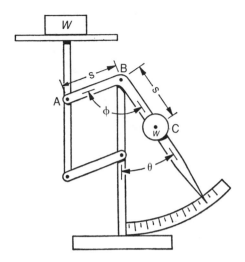

Figure 5.15 Letter balance.

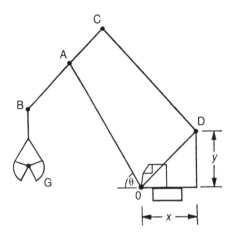

Figure 5.16 Level-luffing crane.

Q.5.4(2). Figure 5.16 shows the mechanism of a level-luffing crane. The jib OA makes an angle θ with the horizontal and is 14 metres long. The jib-head BC is 9.8 metres long and BA is 7 metres. The hoisting wire passes round small pulleys at B, C, and D to the cab. D is fixed relative to the cab and CD is a tie of fixed length. Find a position for D such that as θ is varied without the hoisting wire being reeled in or out, the grab G remains at a nearly constant level ('level luffing'), given that, when θ is 75°, BC makes an angle of 15°, with the vertical. (Make the level of G the same for θ = 45°, 60°, 75°.)

Q.5.5(3). Figure 5.17 shows the cooling curve of a gas which is being liquefied. The effective ambient temperature (heat-rejection temperature) is 300K. The gas is to be liquefied by heat exchange with a single stream of perfect gas heating from T_1 to T_2. Find roughly the best values for T_1 and T_2, other things being equal; ignore the practical need for a small temperature difference at the point P.

Q.5.6(4). Figure 5.18 shows the principle of the Terry 'Anglepoise' lamp. A light arm ABC carrying weight W at A is hinged to a vertical rod BD at B. The linear

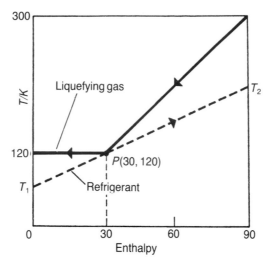

Figure 5.17   Cooling curve for gas liquefier.

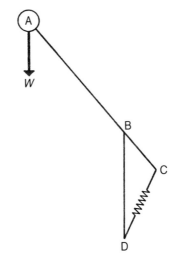

Figure 5.18   Matching principle of Terry 'Anglepoise' lamp.

spring CD is chosen so that it just counterbalances W in all positions, i.e. the arm ABC is in neutral equilibrium. Find the unstretched length of the spring.

Q.5.7(?).   Figure 5.19 shows in plan and aft elevation a typical arrangement of the lead of the mainsheet on an International $10\,m^2$ sailing canoe, a fast single-handed craft. The sheet, which serves to haul in the mainsail attached to the boom, runs from a fixing, A, through a single block, S, attached to the boom and back through a guide block, G, to the helmsman at H, who thus has a constant mechanical advantage of two. The ideal system has a variable mechanical advantage, ranging from less than one, when the boom is right out (dotted line in plan) and the force required at S is small, to about 3 with the sail hauled in hard, when the force required at S is greatest. Devise an arrangement better than that of Figure 5.19 in which the boom can be hauled in faster to begin with and there is less rope lying around when closehauled. Note that the large downward component of the force at S when the boom is hard in is desirable as it pulls the sail into a flatter form. The writer has failed to solve this problem.

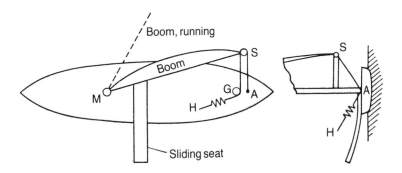

Figure 5.19   Sailing canoe mainsheet arrangement.

Q.5.8(3). Consider the problem of matching the circulating compressors of a gas-cooled reactor at part load. These compressors pump the circulating coolant gases through the reactor and boilers, a load which behaves roughly as a simple orifice, i.e. the pressure drop is proportional to the square of the mass flow: it would be easy to match the compressors at part load by dropping their speed, but the cheapest drive is by synchronous electric motors.

Q.5.9(3). Consider the problem of devising a safe-loading indicator for a vehicle-mounted crane where the safe load varies not only with the angle of the jib to the horizontal, but also the angle it is slewed to relative to the axis of the vehicle. Assume that stability is always more critical than strength.

Q.5.10. In a thick-walled homogeneous cylinder subject to internal pressure, the total circumferential stretch at the inside is slightly greater than that at the outside, since the radial stresses are compressive. Thus the product of the circumferential strain and the radius falls with radius, and so the hoop stress falls rapidly with radius: this means that, if the thickness is comparable with the inner radius, the outer layers of material are relatively ineffectual as long as no yielding occurs. This is a serious problem in the design of pressure vessels for very high pressures. List the means that might be employed to match the stresses better.

Q.5.11. Discuss the choice of gear ratio between the hand of the driver and operating cables of the handbrake of a private car.

Answers

A.5.1. R_1.

A.5.2. Just about: since the flows are equal, with efficiencies of 80 per cent the c for the compressor is about 0.8 that for the turbine. If an impulse turbine is used with a u/c of 0.5 and the compressor has $u/c = 1.0$, this gives a compressor diameter of 1.6 times that of the turbine. The ψ ratio is thus about

$$\frac{300}{40} \times \frac{38}{11} \times \frac{1}{1.6^2} \times 1.25 = 12.7,$$

(see equation 5.2) which is just about possible (cf. Figure 5.2). However, in such a plant the turbine efficiency is much more important, since one unit of heat extracted at 30K takes 9 units of work even in a reversible machine, and probably 15 in a real one. Hence we should design the turbine to be as efficient as possible, and then a single centrifugal impeller may not match.

A.5.3. 4 units, 118°. The sensitivity is

$$W \cdot \frac{d\theta}{dW} = \frac{\sin(\phi - \theta)\sin\theta}{\sin\phi}$$

which is symmetrical about $\theta = \phi/2$; when AB, BC make equal angles with the vertical the sensitivity is a maximum, hence W must then be the geometric mean of its extremes $= \sqrt{1.16} = 4$. The sensitivity when $W = 1$ or 16 then turns out to be

$$\frac{4\sin\phi}{17 + 8\cos\phi}$$

which is found to be a maximum when $\cos\phi = -\frac{8}{17}$, or $\phi = 118°$.

A.5.4. $x \simeq 4.5$ m, $y \simeq 4.9$ m. The jib and jibhead can be drawn in each of the three positions; the distance GB does not change in luffing so B also has the same level in all three. Thus three positions of C are found, and D is the centre of the circle drawn through them.

By drawing velocity diagrams the slopes of the path of B may be found for each of the three positions, and hence the path itself may be sketched very closely.

It is interesting to observe the degrees of freedom of choice that exist for improving this solution. We can vary the angle BAC and the length AC, and also the angle of the jib at which it and the jib head make equal angles with the vertical. We can lead the hoist round a large pulley at C, we can move the pulley away from C slightly, and we can take the hoist to a point other than D. Altogether about eight degrees of freedom can be mustered besides the two co-ordinates of D, which might enable us to make B level in eleven positions. This would be too difficult, however; five points ought to be quite good enough.

A.5.5. Approximately, $T_1 = 75$K, $T_2 = 210$K gives the least matching loss. The loss of available energy of the perfect gas

$$-\Delta b = \int_{T1}^{T2} T_0 ds - dh = \int_{T1}^{T2} dh \left(\frac{T_0}{T} - 1\right)$$

$$= c_p T_0 \ln \frac{T_2}{T_1} - c_p(T_2 - T_1)$$

The mass flow is inversely proportional to $T_2 - T_1$ so we have to minimise

$$\frac{\ln T_2/T_1}{T_2 - T_1} \quad \text{or maximise} \quad \frac{T_2 - T_1}{\ln T_2/T_1}$$

which is a temperature and may be recognised as the 'log mean temperature'. This may be got up to 131K. However, $T_1 = 50$K; $T_2 = 260$K, and $T_1 = 100$K, $T_2 = 160$K involve increases in ideal work of only about 2 per cent.

A.5.6. The unstretched length is zero. Resolve the spring force at C into components parallel to BD and BC. The latter has no moment about B, so the BD component must be constant and equal to $(AB/BC)W$. Regard BCD as a triangle of forces for C in which BD represents the BD component to a certain fixed scale and so CD represents the spring force to the same fixed scale. Therefore the spring force is proportional to CD, and the unstretched length of the spring is zero. The spring is prestressed to that the coils close up tight together when free and need a certain minimum force to start separating them. This force is the stiffness times the free length.

The 'Anglepoise' lamp is a beautiful example of matching. It has two degrees of freedom that require balancing, since another rod is jointed to ABC at A; both are balanced by springs of the type described, both sets of springs being at the base. A parallelogram mechanism carries the balancing force to the upper rod.

A.5.7. No practical answer is known to the writer. The problem would be much easier if the higher mechanical advantage were required with the boom right out: a single stop would then serve where the actual problem requires some sort of active clamp that positively 'takes hold' of another part of the mechanism, what the writer calls an 'active stop'. There is a fundamental law of mechanisms involved [26]. The writer's best suggestion is a windlass with a conical drum at A with a single rope straight to S. The drum is turned by the helmsman pulling a rope wound on another cylindrical drum on the same axle. This axle runs in a fixed coarse pitch nut that shifts it axially to control the feeds on and off the drums. (Why not use a ratchet instead of the cylindrical drum?)

A.5.8. With constant-speed machines the principal matching means available are throttling, bypasses, shutting down some machines out of a number in parallel and variable-angle blading. Speed control can also be had by a fluid coupling with variable filling, which can be outside the shielding.

A.5.9. Two possibilities are:

(1) a load-measuring device at the jib-head with a warning level set by a microprocessor fed with angles of slew and luff, jib extension etc.

(2) a device using load cells built into the underframe.

Note that (2) is not affected by a sloping terrain as (1) is.

A.5.10. Three possible means are:

(a) prestress, i.e. induce an initial tension in the outer layers balanced by a compression in the inner layers. This method is used in gun barrels,

(b) make the cylinder as two (or more) concentric tubes with a small gap between, and admit a fluid at a suitable intermediate pressure to the gap. This is only practical where some sure supply of intermediate pressure is available, as in the HP cylinder of steam turbines. In this case there is the further advantage that the outer cylinder, which does not have to withstand such high temperatures as the inner one, can be of cheaper, stronger material (*cf.* Section 9.2),

(c) make the outer layers of material of higher E (this is not so remote as it sounds).

A.5.11. The chief object must be to keep up the stiffness of the system as 'seen' by the brake pad. A cable may be regarded as a spring of stiffness s, say. If the gear ratio between cable and pad is R, then a force, F, at the pad produces a force F/R in the cable, which stretches it F/sR. The movement at the pad due to this stretch is $1/R$ times as much, or F/sR^2, i.e. the stiffness of the cable 'seen' by the pad is sR^2, so R should be large. The practical limit is set by the inconvenience of the large cable travels which result from a large R, e.g. in the accommodation of the levers used to balance the loads in the cables. Another consideration is that a large reduction R is difficult to arrange at the brake, and it helps in this respect to put part of the overall gear reduction at the driver's end.

6 Disposition

6.1 Problems of disposition

A problem of disposition exists where a limited quantity of some commodity, usually space, has to be shared out to the best advantage between a number of functions. Cases which meet this strict definition are not very common, but cases exhibiting similar features abound, so that the approaches discussed in this chapter are of very wide application.

Some problems of disposition reduce to formal optimisations like those of Chapter 3. Some can be exactly optimised by simple reasoning. Many others can be brought to a practical optimum by such reasoning, once the problem has been suitably framed.

A historic example of the problem of disposition, the commodity being the centre distance of cylinders in an i.c. engine, is given in reference 27.

6.2 Dovetail fixing

Consider the design of a dovetail fixing for small blades in axial flow compressors (Figure 6.1). The female form is to be broached axially through the parallel-sided disc rims and the male form broached or ground on the blade.

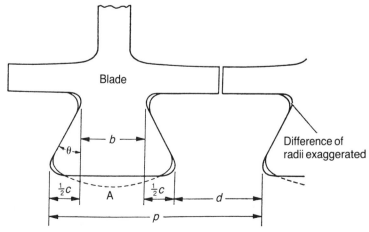

Figure 6.1   Design of dovetail blade fixing.

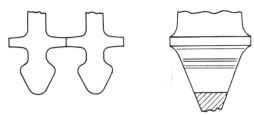

Figure 6.2   Alternative form of dovetail.

For simplicity the materials of blade and disc will be assumed to be metals of similar strength, and the load will be assumed to be direct, without significant bending. The limited commodity is the pitch, p, which has to be shared out between three uses, the blade neck width, b, the overhanging portions of combined width, c, and the disc neck width, d, since

$$p = b + c + d$$

Clearly the division of p should be made so that the load at which all three types of failure (blade neck fracture, disc neck fracture, and shearing or crushing of the lugs) are calculated to occur is the same. For the given conditions, this implies making b equal to d, and c about the same, depending slightly on the material properties (for high tensile aluminium alloy, c would be relatively larger than for ferritic steels: the ratio of c to b and d and the angle θ are best determined by experiment).

This treatment is, of course, superficial. The load designed for should include a vibratory bending term, and there are additional centrifugal loads on the disc neck, and, to a lesser extent, the lugs due to the inertia of the fixing itself. Above all, the form is traditional rather than logical; at the very least there should be some reverse batter on the lugs (dotted line, Figure 6.1). The forces in the disc necks have to flow down into the web of the disc, skirting the area A which mostly has the effect of increasing the stress concentration (on the other hand, it does carry hoop stresses which do some holding together: see Section 3.11). Also, since the load has to be gathered up axially into the thin web, some thinning of the disc profile should start from immediately below the blade (see Figure 6.2).

The fir-tree form (Figure 2.1) cuts the Gordian knot of this disposition problem by overlapping the widths b, c and d so that the only absolute limitations become $b < p$, $d < p$ and joint efficiencies of over 0.8 can be had, as against a limit of about 0.3 for forms with a single pair of lugs.

6.3   Splined shaft

Figure 6.3 shows a section of a gas turbine disc-hub and shaft. The disc is centred and prevented from rotating on the shaft by six splines, flank fitting so that the centring is not affected when the centrifugal stresses cause the disc to grow relative to the shaft (by about 0.003 of the diameter). To keep the shaft stiff, it should be as big as possible: to keep the disc weight down, the hole in it should be as small as possible. Thus cross-section is the commodity we have to share out. Consider the

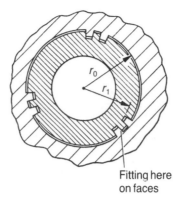

Fitting here
on faces

Figure 6.3   Splined shaft and hub design.

area of cross-section lying in the annular region between the radii r_1 and r_0. Clearly, this cannot be given to either disc or shaft as an unbroken ring. But as broken pieces it is of great value on the shaft, since it adds to the second moment of area of cross-section greatly, being at the maximum radius. As broken pieces it is of no value, or even negative value, on the disc, since it cannot carry any extra hoop stress and long unbroken pieces will cause a greater stress concentration. Thus the right way to design the cross-section is as shown, with a minimum number of shallow grooves in the shaft and shallow narrow splines in the disc bore (torsional stresses are low).

This sort of disposition problem can become very elaborate on stepped splined shafts with several sets of cooling air passages. Notice the important idea of attributing different values to particular portions of the cross-section (the commodity) according to the use to which they are put. We then try, as far as possible, to allocate areas to uses so as to maximise the sum of values (see next examples).

6.4   Disc brake caliper: breaking logical chains

Figure 6.4 shows a disc brake in section. The extreme radius over the caliper is limited to a fixed value R by the requirement to clear the wheel. It has already been

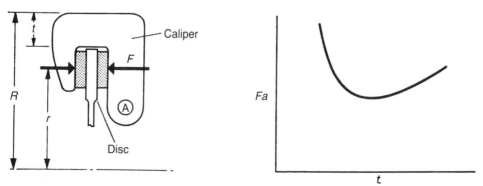

Figure 6.4   Disc-brake caliper disposition problem.

shown that an important measure in this design problem is the *corner work* $F_N d$, where F_N is the nipping force over the pads and *d* is the effective maximum travel of F_N, including *a*, the elastic deflection of the caliper, and the clearances between the pads and the disc.

Now if the radius to the centre of the pads, *r*, is reduced, then *F* must be increased in the same ratio to maintain the same braking torque. The stiffness of the caliper can be increased by increasing the thickness, *t* (Figure 6.4), but at the expense of reducing *r*. As *t* is increased, *F* increases but *a* decreases. The product, *Fa*, varies in some such way as shown in the graph in Figure 6.4. It is possible to find an algebraic expression for *Fa*, quite simply, and to find the minimum by putting $\partial(Fa)/\partial t = 0$. It is better and easier, however, to plot *Fa* against *t* as in Figure 6.4; it gives the information we want, which is not the exact point at which the algebraic expression is a minimum, but the way *Fa* varies with *t* to a practical level of accuracy. This can be used in a further optimisation, say, taking into account material costs.

Notice that the commodity *R* to be disposed of is most valuable in the disc role when it is nearest the outside, whereas its value as caliper thickness *t* is roughly equal everywhere (it is slightly more valuable further in, due to greater curvature of the median surface). The logic of the disposition problem is clear; we ought, other things being equal, to put the disc outside the caliper as in Figure 6.5. Other things are far from equal, however, so we have to return to the original scheme.

We can also regard this excursion, which proves fruitless but might not have been, as an example of breaking the link in a logical chain. The link is that *a* can be reduced by increasing *t* only at the expense of increasing *F*. The weakness in the link is that it is only true if the disc is inside the caliper. It is a very useful practice to build up logical chains fairly rapidly, without, as it were, thoroughly testing each link, and then subsequently to re-examine them (Section 9.5).

Finally, we could regard this as a case of inversion (Section 9.4). All these three kinds of thinking:

   (1) disposition and allocating bits of space to the uses for which they have the most value,
   (2) logical chain making and breaking, and
   (3) inversion, as part of a systematic variation of the configuration,

lead to the same idea, but the first is the best and surest. Notice that in both (1) and

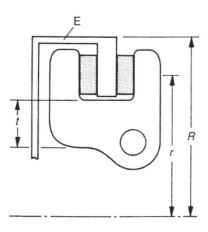

Figure 6.5   'Inverted' form of disc brake.

(2) the idea appears as an improvement, at least in one respect, while in (3) it comes only as one of a large number of random variations and without any *a priori* knowledge of the advantage it possesses. Both (1) and (2) are synthesising steps, but (3) is merely a combinative routine, a mutation.

Notice that the arrangement of Figure 6.5 does not reduce the available space for the hub and wheel-carrier. It is true that a little of R is now taken for a new use, that of accommodating the cylindrical extension E of the disc. On the other hand, F is smaller, the radial extent of the pads is smaller, the arm of the bending moment which F exerts on the thickness t is smaller, and these effects should reduce the radial extent of the entire caliper more than enough to offset the cylindrical extension. Indeed, when the position of the pin A is considered, the space restrictions in the centre should be eased (Section 7.2).

The new arrangement has at least three disadvantages.

(1) the removal of dirt and large flows of water from the 'disc-drum' presents problems,
(2) the 'disc-drum' is expensive, and
(3) the axial extent is increased.

These are probably enough to eliminate it.

6.5   Alternator rotor

Consider the section of an alternator rotor shown in Figure 6.6. It is simply a rotating electromagnet with two poles, energised by a coil wound round the magnetic axis in slots cut in the rotor. The centrifugal loads on the rotor tend to burst it, and this limits the diameter, since the speed is fixed at 3000 rev/min by the need to generate at 50 Hz (cycles per second). Different designs of rotor will have different masses and strengths, and so slightly different limiting sizes, but for the moment the diameter will be taken as fixed. Cross-sectional area is the commodity we have to dispose of to best effect, and it must be shared between the following functions:

s strength, to resist bursting,
f flux path, for the magnetic field,
c conductors, for the exciting current,
i insulation, and
k cooling (by water or hydrogen, say).

Not all these functions are required in all areas equally, and in Figure 6.6 the regions in which various groups are most important have been shown by broken lines—broken because the boundaries are indistinct. All five functions are competing for space in the slotted regions, a section of which is shown in Figure 6.7. It has the conventional parallel slots, giving a 'tooth' which is wider at the tip than the root. However, there is a flux 'bottle-neck' at the narrowest part of the tooth, C, and so the triangular areas shown shaded are wasted as far as function f is concerned—the flux through the tooth is limited (practically speaking) to the saturation flux through C, and could almost equally well be carried by a parallel tooth of the same root width, as shown by the broken line.

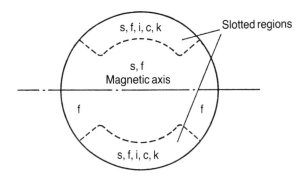

Figure 6.6 Alternator rotor (1).

As far as function, s, is concerned, the shaded areas are again wasted, for the centrifugal load is a maximum at C, and less at wider sections such as A and B. We conclude therefore that it is the *teeth* that should be parallel, not the slots. True, there is a problem of transferring the centrifugal loads on the conductors into the teeth, but this merely complicates the argument without invalidating the conclusion.

AEI made a change in this direction in 1968 in their 660 MW design. Instead of using a parallel slot as in their previous alternators, they introduced a slot tapered at the bottom but parallel at the top, giving a gain in conductor area of 16 per cent. While this is less than a slot tapered all the way would give, it leaves a little extra metal on the sides of the tooth tip in which to cut the wedge grooves which retain the conductors.

As in all such cases, there arises the question of how to distribute the fresh supply of cake—AEI increased the ampere turns by only 8 per cent, using the other 8 per cent to reduce the current density, keeping the rotor copper losses unchanged though the rating of the alternator per meter of length was greater.

6.6 Structures: the feather

In a rather broad sense, much structural design is concerned with disposition, how to use a limited amount of material to give the most strength. Moreover, when the

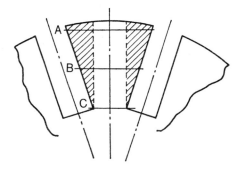

Figure 6.7 Alternator rotor (2).

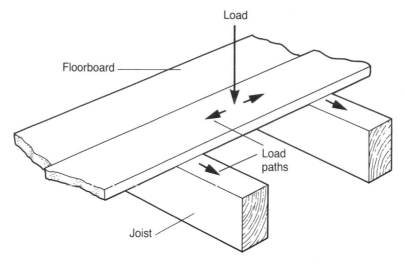

Figure 6.8   Floorboards and joists—a tiered structure.

space available is limited, another constraint is introduced. Both these aspects are exhibited by that very interesting natural example of structural design, the feather.

Where relatively small loads have to be carried over a large area, as in a feather or a floor, it is usual to use a 'tiered' structure. In a floor, floor boards support your feet and carry the load to joists, which in turn carry the load to the walls (Figure 6.8). Where the span is larger, the second tier of wooden joists may be supported in turn by a third tier of steel joists. Each tier lies at right angles to the tier before, and carries the load from the tier before to the tier after.

A bird's wing has four tiers, of which the quill or rachis—the central member of the feather (Figure 6.9)—is the third. The air loading is borne first by tiny barbules which carry it to the barbs which spring from either side of the rachis. From the barbs the load is carried to the rachis and finally to the bones and muscles of the wing. The section of the rachis is circular at its insertion in the wing (the part used as a pen nib in the past) but rectangular in the part with the barbs (Figure 6.10). Most of the material is in the 'flanges' (see Figure 6.10) as might expected, while the webs are very thin and would be unstable but for the fact that the interior space is full of a

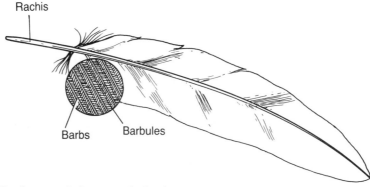

Figure 6.9   Structural elements of a feather.

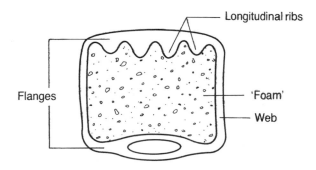

Figure 6.10   Section through rachis.

foam-like structure of keratin (the material of the feather). This 'foam' prevents the thin shear webs from buckling.

The most interesting point is the upper flange. Because the feather is bent upwards by the air loading in flight, it is in compression, and so in danger of buckling. It is stabilised by integral longitudinal ribs. Now it would be possible to stabilise the flange by transverse ribs, but then the rib material would not contribute to the flange area and hence to the bending strength. The lower flange is in tension, and so inherently stable and not in need of stiffeners.

Ships, aircraft and bridges

Wooden ships were constructed of ribs or frames at close spacings crossed by longitudinal planking, a two-tier structure like a wooden floor. Iron and steel ships were at first built in a similar style, with mainly transverse frames and ribs, but for the same reason as in the feather, more longitudinal stiffening is now used. A similar philosophy is followed in aircraft wings and in box girder bridges. Ships' hulls, wings and bridges are all essentially beams and the principle is to dispose the material, as far as possible, in such a way that it provides both stiffening and flange sectional area, rather than just stiffening.

A further point to note about the feather and the aircraft wing, which does not apply to the other cases, is that there is a constraint on depth. For aerodynamic reasons, the thickness through wings and feathers needs to be kept small, whereas resisting bending would be made easier by greater depth. In vehicle design there are similar constraints on where structural material may be placed, arising from the need for good all-round vision or the need to avoid encroaching on the seating space.

6.7   Structures: form design

The feather and related designs by man touch on a more generalised view of disposition, that of the disposition of material in space, or form design. This is a key area in the embodiment stage, but one which must also often be considered at the concept stage as well. In form design it is usual to have to reconcile the conflicting demands of structural strength and economy in manufacture, and sometimes also

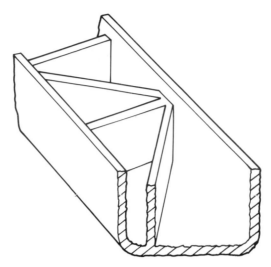

Figure 6.11   Shaft of a toy spade.

those of fluid mechanics, ergonomics, appearance and maintenance. It is a very
large and difficult subject, hardly researched at all and presenting great intellectual
demands.

Consider a toy spade for use on the beach. The most suitable material is a
polymer and the appropriate method of manufacture is injection moulding, but
there is a conflict in the form design between structural and moulding
requirements. The shaft has to withstand bending and twisting, and the ideal form
for this would be tubular: however, a tubular form is difficult to mould. A channel
form presents no difficulty in moulding and is good in bending, but twists very
easily. The form shown in Figure 6.11 is commonly used: it combines adequate
torsional stiffness with suitability for moulding.

Figure 6.12 shows a case of manufacture using polymers, a solar panel in which
the collecting surfaces are blackened fins on copper pipes connected by manifolds.
The collecting elements are encased by glass tubes, which are connected by
polymer casings at each end in rigid panels of six. The casings are split on a plane as
shown in the figure. Two lower halves are located on dowels on a table, six glass
tubes and collecting elements are placed in them and the manifolds assembled to
them, the casing parts acting as a jig. The top halves are then added and foam is
injected into the hollow casings. Some foam squeezes past the glass tubes into
circumferential grooves in the casing, becoming denser in the process and sealing
the joint. The result of this simple operation is a rigid panel, with the copper
elements positively located and thermally insulated in the non-collecting part, and
with the glass-to-casing joint sealed.

Figure 6.13(a) shows a common way of using a helical coil spring in a vehicle
suspension, in which the end of the spring has been ground off flat and bears on a
flat surface on the suspension arm. Figure 6.13(b) shows an alternative approach in
which the spring is not ground flat but left as coiled, and a depression is formed in
the suspension arm into which this unground spring end fits. For a small extra cost
in the press tool, an operation is saved on the spring, an operation which could
impair its strength.

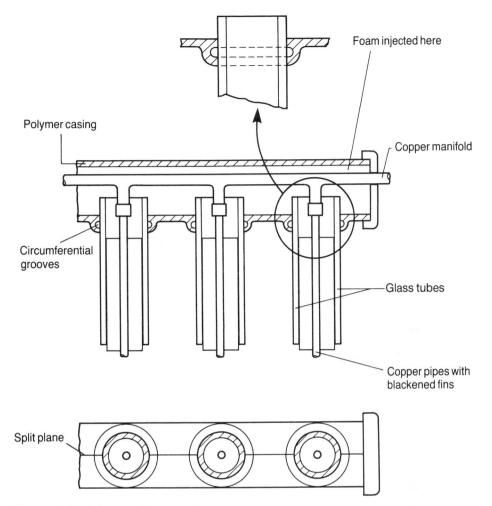

Foam injected here

Polymer casing

Copper manifold

Circumferential
grooves

Glass tubes

Copper pipes with
blackened fins

Split plane

Figure 6.12 Solar panel construction.

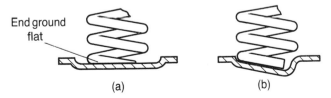

End ground
flat

(a)

(b)

Figure 6.13 Suspension springs.

6.8 Joints

Finally, the problem of making a longitudinal joint in a pressurised cylinder will
serve to illustrate an important principle of form design. The obvious solution is
shown in Figure 6.14(a), with two fairly shallow flanges bolted together. The
pressure in the cylinder produces tensions, T, in the joint, which result in severe
bending because of the large overhang, h. A much better design is shown in Figure

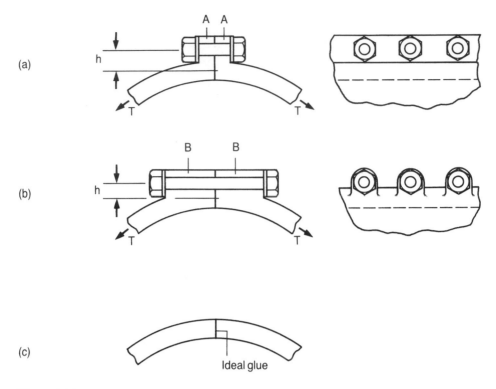

Figure 6.14   Longitudinal joint in a pressurised cylinder.

6.14(b), where the 'flanges' are much thicker but are reduced to local bosses, B, so there is no significant change in material quantity. The overhang, h, is very much less, and the bending moments are reduced in the same proportion. Moreover, the sections resisting those moments are deeper and so stronger for the same cross-sectional area.

A principle of form design

The principle of form design involved may be stated as follows. A well designed joint should approximate as closely as possible to the continuous form which would be adopted if no joint were needed, i.e. in the case of the cylinder, an unbroken cylinder. Form (b) meets this requirement much better than form (a).

  One way of looking at this principle is to imagine an ideal joint, consisting of a simple cut through an unbroken cylinder (or other shape) which is then stuck together again with an 'ideal glue' as strong as the cut material but capable of being dissolved at will (Figure 6.14(c)).

6.9   The design of gear teeth

In what follows it will be assumed the reader has some knowledge of involute gearing. The specific problem that will be treated is the design of ground involute spur teeth for a pair of gears of given diametral pitch and numbers of teeth. The

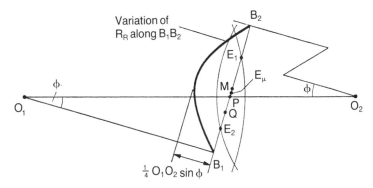

Figure 6.15   Design of involute gearing.

restrictions that may be imposed by standard cutters will be ignored, except that a standard base circle pitch will be assumed. The centre distance will be regarded as variable within the small range necessary to exploit the available degrees of freedom.

This is a problem of disposition, in so far as we aim to make the load capacity per unit face width as large as possible. Like the form design problems of the previous section, it illustrates the wider class of problems that have only some of the characteristics of disposition.

Since the numbers of teeth and the base circle pitch are fixed, the base circle radii O_1B_1 and O_2B_2 are fixed (Figure 6.15). By increasing or decreasing O_1O_2, the centre distance, the pressure angle ϕ may be varied. We can also choose the tip circle radii O_1E_1 and O_2E_2 to suit ourselves. Finally, we can choose the chordal thickness of one set of teeth at some suitable radius: the thickness of the other set will then be fixed by the need to have a suitable backlash.

We have thus four degrees of freedom of choice—ϕ, two tip circle radii and one tooth thickness. To simplify the problem, we shall suspend one of these by making the tip circles overlap by two modules (2/diametral pitch).

Suppose Q is a point of contact (Figure 6.15), then the radii of curvature of the flanks of the teeth at Q are B_1Q and B_2Q and the relative radius of curvature is

$$R_R = \frac{B_1Q \cdot B_2Q}{B_1Q + B_2Q} = \frac{B_1Q \cdot B_2Q}{B_1B_2} \tag{6.1}$$

where B_1, B_2 are the points of tangency to the base circles.

It was seen in Section 4.8 that the load-carrying capacity of a line contact such as that between gear teeth might be expected to be proportional to R_R. The largest value of R_R occurs when contact is at M, the mid-point of B_1B_2, and is $\frac{1}{4} B_1B_2$ or $\frac{1}{4}O_1O_2 \sin \phi$. The variation of R_R along B_1B_2 is plotted in Figure 6.15; its form is parabolic, and symmetrical about M with zeros at B_1 and B_2.

The three principal sorts of failure we have to avoid, in the absence of shock loads, are:

(1) surface fatigue failure,
(2) 'bending' fatigue failures at the roots,
(3) scuffing failure.

To make the argument tidier, we shall ignore (3) for the moment, and see how the exercise of our three degrees of freedom of choice affects (1) and (2).

(1)   Surface strength

Two of our three degrees of freedom of choice affect surface strength, ϕ, and whichever tip circle radius we regard as the independent variable, say, O_1E_1 (since $O_2E_2 = O_1O_2 + 2$ modules $- O_1E_1$).

For any point dividing B_1B_2 in a fixed radio, R_R is proportional to $\sin \phi$, so that increasing ϕ increases the load capacity of a single tooth contact. As there will be a single pair of teeth in contact at and a little on either side of E_μ, the mid-point of E_1E_2, we might propose as a figure of merit for surface load capacity

$$(R_R)_{min} \times \cos \phi \tag{6.2}$$

where the $\cos \phi$ factor allows for the higher tooth load produced by a more oblique pressure line and 'min' means minimum for a single pair of teeth in contact. To make $(R_R)_{min}$ as large as possible we make E_μ fall on M, the mid-point of B_1B_2. Then $(R_R)_{min}$ is virtually equal to R_R at M (perhaps 1.5 per cent less at most) so that the figure of merit becomes $\frac{1}{4}O_1O_2 \sin \phi \cos \phi$. We therefore make ϕ as large as possible.

The improvement to be had in the figure of merit compared with 'standard' 20° pressure angle gears by these two means, increasing ϕ and moving E_μ to, or near to, M, is of the order of 80 per cent in a typical case. The writer believes that the 'single pair' form of analysis given here underestimates the value of the higher contact ratio of lower pressure angles in the case of high-speed gears, where the assumption of constant wind-up in the teeth may be nearer the truth than that of constant tooth force. Nevertheless, the improvements in surface strength are at least of this order [28], perhaps because the effect of R_R is greater than Hertzian theory indicates.

(2)   Bending strength

Figure 6.16 shows the effects of increasing the pressure angle on the bending strength of teeth. The effect of increased ϕ is

     (1) to increase the tooth force as $\sec \phi$,
     (2) to increase the radial component of tooth force,
     (3) to decrease the bending arm h of the tooth force, about the mid-point of the base AB, and
     (4) to increase the width of the base AB.

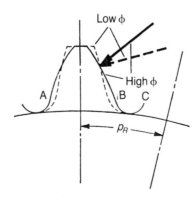

Figure 6.16   Effects of change of pressure angle.

All but (1) are advantageous, and (1) is small and partly cancelled by (2). The dominant effects are (3) and (4), which greatly increase the bending strength. Again the indication is to increase ϕ, but here the necessary limit on the process is apparent (Section 4.6). As ϕ is increased, both the lands at the tips of the teeth and the spaces between their roots become narrower, and both these effects must be watched. In particular, the root space presents a disposition problem in itself. Suppose p_R is the circular pitch of the teeth a little above the root radius. This has to be shared between the tooth base AB and the root space BC (Figure 6.16), and there will be an optimum ratio of division. If the root space is too narrow, the small fillet radius will more than offset the wider tooth base: the writer has used a ratio of 3:1.

In 'standard' gears, p_R is less for the pinion than for the wheel, since the circular pitches are equal at pitch circles and the centre lines of the teeth converge towards the centre more rapidly on the pinion. By shifting E_μ from the pitch point P towards M, the bending strengths of the wheel and pinion can be balanced. To some extent balance of bending strength can also be effected by varying tooth thickness, but clearly this cannot replace a sheer deficiency of p_R in the pinion. We can regard the whole bending-strength problem of the gears as one of disposing the combined p_Rs between two tooth bases and two tooth spaces, except that as we move E_μ from P towards M the sum of the p_Rs increases.

Scuffing strength

It is even less possible to be accurately quantitative here than in the other cases, but the aims must be to keep down the surface stress and the sliding velocity. Since the sliding velocity at any point of contact Q is $(\omega_1 + \omega_2)PQ$, where ω_1 and ω_2 are the angular velocities of the pinion and wheel, contact must be confined to a region as close to the pitch point as possible. The first of our two devices for improving surface strength, increasing ϕ, reduces the sliding velocities also, but the second, shifting E_μ to or towards M, increases the maximum sliding velocity.

We can summarise our findings with regard to the most important two of our three degrees of freedom as follows:

Kind of strength	Indications	
	Pressure angle ϕ	Position of E_μ
Surface	Increase as much as possible	At M, or as near M as possible
'Bending'	Increase to limit imposed by disposition of p_R or 'sharpening' of tips	On wheel side of pitch point P, sufficiently to balance bending strengths of pinion and wheel
Scuffing	Increase as much as possible	P

Non-involute gears

It is possible to have tooth forms other than involute, with inherently higher surface strength. On a Hertzian basis it can be shown rigorously that the practical limit is almost reached by gears with circular arc profiles (Wildhaber, Novikov) [29].

6.10   Disposition of allowable stress

It often happens that the maximum stress in a component is the sum of stresses arising from several causes. In the suspension-bridge tower of Section 3.4, the maximum stress is due to the combined effects of the downward load applied by the cables, the bending moment caused by the wind, lack of concentricity of the cable load and the axis of the tower, and possibly the fixing moments of the transverse members where a portal form of structure is used (see Question 10.11).

In such cases it may be appropriate to approach the design problem as one of disposition, where the allowable stress is to be shared out among the different terms in the most favourable way possible. The final criterion of the best design will be overall cost, but often some intermediate criterion, such as quantity of material used, can be substituted to make the problem more manageable.

In most of these problems the severest stresses due to each cause will not occur simultaneously, or fatigue may be concerned together with a steady loading, and then simple addition of stresses will not do. Special techniques are then needed, such as a Goodman diagram or a statistical treatment, in order to decide what are permissible combinations of loadings.

The simplest case is when two stresses, arising from different causes and varying in different ways with the scantlings, add directly at some points. Consider a strap which is both bent and loaded in tension, and suppose the bending to be due to an imposed *distortion*, i.e. the strap is forced to deflect a given amount. Then the bending stress will vary directly as the thickness, t. On the other hand, if the direct stress is due to an imposed *load*, it will be inversely proportional to t, and the combined stress will be of the form

$$At + \frac{B}{t} \tag{6.3}$$

where A and B are constants and the first term is due to bending and the second term to the direct load. Question 7.2 is of this sort, and an example is given on page 6 of reference 30.

6.11   Electromagnets

The design of electromagnets to produce intense magnetic fields involves some interesting disposition problems. Figure 6.17(a) shows the hollow cylindrical coil of an air-cored solenoid. The portions of the coil nearest the axis produce more ampere turns per metre of conductor, so that an optimum distribution of current density is non-uniform (in fact, inversely proportional to the radius). This can be achieved by making the coil of a thin flat spiral conductor [31] as in Figure 6.17(b).

In an electromagnet using iron, the effect of the iron can be considered as that of minute magnets, or dipoles, lying along the magnetic field. In Figure 6.17(c), if O is the point at which we want to produce a strong field in the direction H_R, and A is any point in the iron, we can imagine a dipole at A producing a small increment of field at O, which will have a component along H_R. If we rotate the dipole in the plane of the figure, this component will vary, and it makes its maximum contribution in

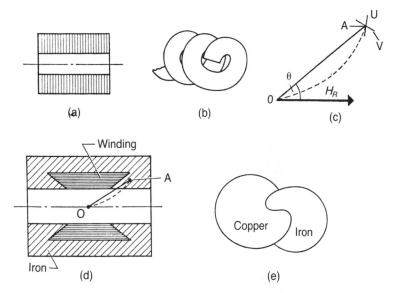

Figure 6.17 Electromagnets.

position U. In another position, V, it has no component, and when turned through 180° it opposes H_R. Thus we seek to arrange all the dipoles in the iron in the U position: difficult as this task may sound, the design by Bitter in Figure 6.17(d) achieves it. The working volume of the field is produced in the central hole in the iron, at O. The coil should have the same current density variation as before, i.e. inversely proportional to radius, but this time it is achieved by winding the coil in a single spiral from a tapered strip of foil, beginning in the centre with the narrow end. This yields the cylindrical coil with conical depressions in the ends shows in the figure.

Figure 6.17(e), which follows Laithwaite, is an idealisation of an electro-magnetic device—motor, magnet, choke—as a magnetic circuit linked with an electric circuit as closely as possible. Laithwaite [32] has an important concept—his 'goodness factor', which is a measure of the excellence of the linking of the two circuits. Goodness factor is an example of the sort of idea which springs only from design, and is so lacking in the engineering vocabulary (see Section 4.5).

Questions

Q.6.1. Consider the design of a hotel bedroom to accommodate two single beds 0.9 m × 1.9 m, a wardrobe with drawers inside and double-hinged doors, 1.2 m × 0.6 m in plan, a vanitory unit 0.5 m × 1.2 m, two wicker armchairs, a table 0.6 m × 0.8 m and a luggage stand 0.7 m × 0.5 m. A window is to be provided in one short wall and a door in the opposite wall and the length of the room should be 1.5 times the breadth. There is a radiator below the window. Using sensible guesses where properly you should use ergonomic data, find the smallest size of room you would recommend, and show the arrangement you propose. (A room much deeper than 1.5 times its window wall length will seem gloomy because of poor penetration of

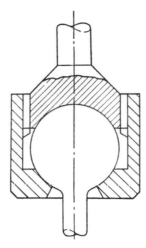

Figure 6.18   Ball-joint.

daylight, unless the ceiling is excessively high, and a squarer room will give a shallower and hence less economical building.)

Q.6.2. Consider the disposition problem set by the design of cupboards for kitchens, especially by the range of heights of the articles stored, which should be assumed to be up to 320 mm, and the need for easy access to most items.

Q.6.3(2). A spur pinion has a circular pitch at the roots of the teeth p_R and 'semi-circular' fillets of radius r between the teeth. If the tooth chordal width at the root is b, so that $b + 2r = p_R$, find approximately the optimum value of r assuming that the bending stress in the tooth may be estimated by the usual beam theory, with a stress concentration factor K given by

$$K = 1 + 0.15 \frac{b}{r}$$

Q.6.4(3). Figure 6.18 shows a sketch for a ball joint connecting two rods in a linkage. Regarding the design of such a joint as a problem of disposition of an overall diameter that is to be kept as small as possible, analyse the functions between which that diameter is to be shared.

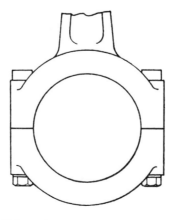

Figure 6.19   Connecting-rod big-end.

Q.6.5. Consider the design of joint faces for the halves of connecting-rod big ends (see Figure 6.19 for an example) as a disposition problem. Make a functional analysis of the joint face, and examine the relative values of different regions of the available cross-sectional area for different uses.

Q.6.6. A heat engine working on the Rankine cycle and using a halogenated hydrocarbon as a working fluid extracts heat from a plentiful supply of hot water at T_1 and rejects it to the atmosphere at T_0. Analyse the functions to be performed by $T_1 - T_0$ and state the disadvantage incurred by reducing the fraction of $T_1 - T_0$ allotted to each.

Answers

A.6.1. About $3\,\mathrm{m} \times 4.5\,\mathrm{m}$. Some general principles are, that spaces for bed-making, swing of wardrobe and room doors, etc. should so far as possible have more than one function, and that one large squarish clear space is much more convenient and comfortable than an equal area disposed in long narrow strips. It follows that the perimeter of the free space should be as small as possible relative to its area—in engineering jargon, the mean hydraulic diameter should be large. This short perimeter must therefore consist largely of bedsides, wardrobe and room doors, vanitory unit, etc.

A.6.2. Access to and removal of most items without disturbing the others is desired. Dividing the accommodation between shelves in the cupboard and receptacles on the back of the door, as in refrigerators, is a great help, but this is limited by the acceptable inertia and weight on the hinges. Shallow drawers suspended under part of the width of a shelf also help, as do intermediate shelves at some levels extending perhaps one-third of the way from the back of the cupboard.

A.6.3. The factored 'bending' stress is roughly proportional to

$$\frac{1 + 0.15\,b/r}{b^2}$$

since the section modulus of the root section of the tooth is proportional to b^2. Substituting $r = \frac{1}{2}(p_R - b)$ (or using Langrange) and minimising (Section 3.10) gives $b = 0.76\,p_R$ approximately.

This result is reasonable, but data are available for better treatments. Furthermore, complications of tolerances, preformed roots, etc. shift the optimum value of b downwards slightly.

A.6.4. The functions amongst which the overall diameter is distributed are (Figure 6.20(a)).

A	ball rod		D	socket end structure
B	freedom		E	joint
C	tensile bearing area		F	tensile structure

It is clear that D need not encroach on the diameter but the others must (except that E can be got round by extreme constructions such as a longitudinally split socket bolted through above and below the ball). It is also plain that it is better to use ball joints in compression where possible. In the front suspension of cars where the stub-axles are carried on vertical members (king pins) with ball joints top and

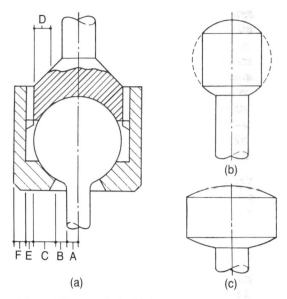

D

FE C B A

(a)

(b)

(c)

Figure 6.20   Disposition of diameter in ball joint.

bottom, these members are sometimes joggled at the lower end so that all the joints
have the stem of the ball downwards. In this case the reduced frictional resistance to
steering in the load-carrying bottom joints is important.

In a proper study, it would be desirable to add functions G (tolerances) and H
(fillets and chamfers). If the freedom required is small enough, it may be worth
using a slightly larger ball turned down round the equator (Figure 6.20(b)) or even
the form (c) if a unique centre of hinging is not necessary.
A.6.5. The functions of the joint are:

(a) transmitting direct load,
(b) transmitting shear load,
(c) transmitting bending moment,
(d) ensuring accurate positioning of one part relative to the other, i.e.
providing a register.

Of these, (c) is the most exacting and (b) and (d) are intimately connected. The
fulfilment of (a) and (c) by a joint of this sort depends on loading the faces together
by a prestressed bolt or bolts, the function of which is that of a very stiff spring. For
this reason, it is better in disposing the various functions in the available area to
forget (a) and substitute (e), providing prestress. Now study the suitability of
various parts of the available cross-sectional area (Figure 6.21(a)). The area is
limited in the directions A and B by clearance problems, but it will make sense to
use the whole rectangle. Since bending occurs in both directions in the plane of the
connecting rod, the prestressing material is best put in the middle and the bending
function given to the 'flange' region, as in Figure 6.21(d). The bolt(s) will thus be
put through the middle of the area. Coming to the shear-resistant and register
functions, any part of the area will do. The stepped face of Figure 6.21(b) is worth
consideration, provided that there is a clearance at C to ensure contact on the areas
F which are well suited to take the bending. The shear-resistant function is
adequately filled, but the maintenance of a register demands making the faces E

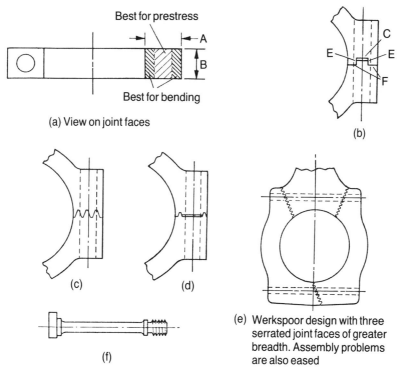

Figure 6.21   Joint faces of connecting rods.

and F all at once. A lighter, more solid job is given by serrations as in Figure 6.21(c). Finally, consideration might be given to removing the central serrations as in Figure 6.21(d), since the shear and register functions are adequately supplied by those remaining and the section stiffness in bending is scarcely altered. On the other hand, the same bolt at the same stress will give a higher prestress in the vital area.

Note that, since the radius of gyration of the cross-section of the bolt is bound to be less than that of the joint face, it is best to free the bolt of bending and increase the preload—we aim at a strict separation of function, the bolt providing a direct prestress only and not contributing in its own right to the bending section at all. To free the bolt of bending stress, we make it a clearance fit. Also since we have decided the ideal function of the bolt is that of a stiff spring, we give it a long waisted shank, as in Figure 6.21(f).

Finally, consider the possibilities of easing the problem by increasing the section A × B of Figure 6.21(a). Increasing B means increasing the cylinder pitch or else thinning the crankshaft webs—neither being desirable. In any case, since we are concerned with beam strength, an increase in A is more to be desired. This can be done without adding to the clearance problems by skewing the joint face to an angle. An even larger gain can be had by the arrangement of Figure 6.21(e), adopted by Werkspoor, at the cost of an additional joint face.

A.6.6. The temperature difference $T_1 - T_0$ has to be divided between

(a) temperature drop in the water,
(b) temperature difference between the water and the working fluid in the boiler,
(c) difference between boiler temperature and condenser temperature,

(d) temperature difference between the air and the working fluid in the condenser, and

(e) temperature rise in the air.

Reducing (a) and (b) (which partly overlap) requires an increased boiler size. Decreasing (c) reduces the Rankine cycle efficiency. Decreasing (d) and (e) (which partly overlap) means increasing condenser size and possibly the power of a fan to circulate cooling air.

Note that since the air side of the condenser is likely to have the smallest heat transfer per unit area for a given temperature difference, we must expect (d) to be greater than (b) in an optimised design.

7 Kinematic and elastic design

7.1 Introduction

While it is not the purpose of this book to discuss special items in the engineering repertoire, the complementary ideas of kinematic and elastic design are so important as to warrant an exception.

The principle of kinematic design, which was expounded and used by Kelvin, may be stated as follows: 'When locating or guiding one body relative to another, use the minimum number of constraints.'

A simple and well known example is that of the three-legged stool, which obeys the principle and always stands firm on its minimum number of three legs. A four-legged chair, however, has one more than the minimum number of constraints, and so will often stand on only three of its four legs. However, if the floor is flat and we adjust the length of the legs so that their ends lie very nearly in a plane, it will rock only slightly. If now we sit on the chair, the stresses our weight sets up will deform it so that all four legs are carrying some load. The four-legged chair is an elastic design, the load distribution in it depending on its flexibility and also on the perfection of its 'fit' to the floor, i.e. its freedom or otherwise from rock when under no load.

7.2 Spatial degrees of freedom: examples of kinematic design

For those who never were, or no longer are, familiar with the idea, some explanation of spatial degrees of freedom may be helpful.

An object has n spatial degrees of freedom if n co-ordinates are required to specify its position. Thus, a moving point in space has three degrees of freedom since it needs three co-ordinates, x, y, and z, say, to specify its position. A line AB of fixed length has five degrees of freedom; A and B would have three degrees of freedom each, corresponding to the co-ordinates $x_A, y_A, z_A, x_B, y_B, z_B$, but because the length AB is fixed,

$$(x_A - x_B)^2 + (y_A - y_B)^2 + (z_A - z_B)^2 = (AB)^2 = \text{constant}$$

and so, if five co-ordinates are given, the sixth is fixed. Another way of seeing this is as follows; if A is fixed, B moves on a sphere, where its position can be specified by only two co-ordinates, say latitude and longitude. Hence the degrees of freedom

153

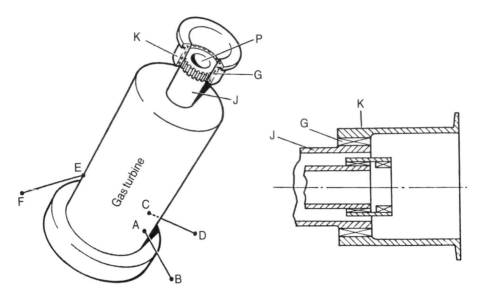

Figure 7.1   Kinematic mounting of helicopter engine.

are three for A, plus two for the latitude and longitude of B. The two points A and B have their degrees of freedom reduced from six to five when the distance between them is fixed. We can express this by saying that a link of fixed length imposes one simple constraint.

If a point is constrained to move on a line, where its position can be specified by a single co-ordinate such as the distance from one end, it has lost two degrees of freedom. Such a constraint is equal to two simple constraints.

A rigid triangle ABC has six degrees of freedom (three for each corner less three simple constraints, the constant lengths of the sides). A corollary is that any rigid body has six degrees of freedom, so that to support an engine in a helicopter, say, six simple constraints are the least which will suffice. Figure 7.1 shows such a scheme diagrammatically.

A helicopter engine mounting

The engine is supported by three struts, AB, CD and EF, freely pivoted at their ends, providing three simple constraints. At the upper end of the engine, there is a tubular extension, J, surrounding the drive shaft, P. This extension slides in a tubular mounting, K, secured to a gearbox which is itself secured to the airframe. J is constrained to move on a line, the centreline of K, constituting two further simple constraints. Also J and K are provided with interlocking gear teeth, so that J is prevented from rotating: this provides the sixth constraint.

Notice that with this arrangement strains in the airframe do not manifest themselves in the engine, and vice versa. The loads in the struts and at K are statically determinate. An incidental advantage is that the path of the torque reaction, from the engine reduction gear casing to the toothed ring K on the aircraft, is the shortest practicable. Installation is simple—the engine is 'plugged' in at K and pins inserted at A, C, and E. The loads at A, C, and E are direct forces—no moments are exerted on the casing at these points.

Added freedoms

The helicopter engine mounting is an example of kinematic design as just defined. There is a very similar kind of application that does not, however, always meet the condition of minimum constraints. It arises when a degree of freedom is deliberately added to a system to ensure that a desired balance of loads is obtained. For example, in the disc brake of Figure 6.4 it is highly desirable that the nipping force, F, between disc and pad should be the same on each side of the disc. This can be arranged very simply by giving either the disc or the caliper a degree of freedom in a direction substantially parallel to the axis of the stub shaft. In Figure 6.4 the caliper is given a balancing degree of freedom by being pivoted about the pin A, which also forms the attachment to the wheel-carrier.

Further examples of load-balancing degrees of freedom occur in those surface plates which stand upon nested series of three-legged members such that all the points of support carry equal loads, and in the support system for some tilting pad thrust bearings (Figure 7.2(a)). If friction and slight variations in geometry are ignored, the levers A will equalise the loads on all the pads.

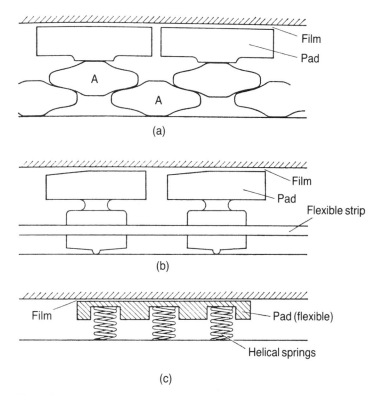

Figure 7.2   Backing of tilting pad bearings.

Epicyclic gearing

An interesting case is that of epicyclic gearing. First consider a simple epicyclic gear, like that in Figure 7.3, as a plane static mechanism. The sun pinion is subject to three tooth loads P_1, P_2, and P_3. If it is left free to float in its own plane, it moves until P_1, P_2, and P_3 are equal, as in the triangle of forces. Balanced load sharing between the three planets can also be achieved by letting the annulus float.

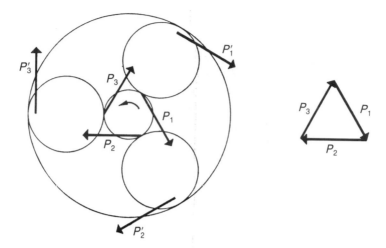

Figure 7.3   Tooth loads in epicyclic gear.

In the dynamic case, of course, the inertia of the sun, annulus and planets means that large alternating unbalanced loads can occur. Here the elasticity of the annulus gear may help, since the deflection under load can be of the order of 0.0015 of the radius.

Let us now consider the epicyclic gear as a three-dimensional device; imagine it being used as a reduction gear with the input to the sun pinion and the output from the spider. The spider has two end plates that support the ends of the three pins on which the planets run, and these three pins alternate with three columns that unite the end plates into a single structure (Figure 7.4).

The tooth loads on each planet are reacted by two roughly equal forces, R, at the ends of its pin. The three forces R at the end remote from the output subject the spider to a torque $3aR$, where a is the pitch circle radius of the planet pins. With practical materials and construction the effect is to twist the spider, so that the pins lie in a slightly skewed position, and the planet gear teeth make an angle with those of the sun pinion and the annulus gear. The practical results are heavy, expensive spiders and badly distributed loading on the teeth.

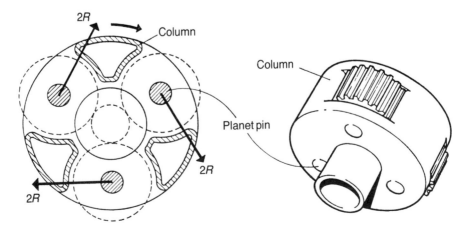

Figure 7.4   Twisting of spider of epicyclic gear.

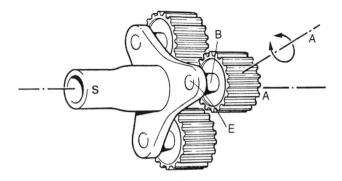

Figure 7.5   Kinematic design of spider and planets.

The solution to this problem is to introduce a further degree of freedom of the planet gears, so that they are free to align themselves with the sun and annulus. The arrangement of such a spider is shown diagrammatically in Figure 7.5. The output shaft, S, carries a single, relatively light, three-legged member, E, bearing three overhung pins B. The planets are pivoted to these pins in such a way that they are free to turn about an axis AA. (This probably implies they have another extra degree of freedom also, about an axis perpendicular to both AA and the axis of B.)

It is instructive to list the items from the engineering repertoire that would be suitable to provide this additional degree of freedom. They can be classified according to whether the degree of freedom is kinematic (i.e. based on relative sliding or rolling) or elastic.

Kinematic

(1) Self-aligning antifriction bearings, where the rolling elements provide for both the rotation of the planet gear and its alignment—spherical roller bearings or self-aligning ball bearings. The load capacity of the latter would normally be inadequate. (Why? see Section 4.9.)

(2) Separate antifriction bearings providing a single degree of freedom. These could not normally go inside the pin, for then they would be too small to carry the load; the alternative is a gimbal or yoke arrangement, which is likely to prove clumsy.

(3) Spherical hydrodynamic (or hydrostatic) bearings providing for both planet rotation and planet alignment.

(4) Spherical (or cylindrical) hydrostatic bearings allowing for planet alignment only.

(Notice that stationary, non-hydrostatic plain bearings, allowing only for alignment, would give too much Coulomb friction.)

Elastic

(1) Pins,
(2) Straps,
(3) Diaphragms,
and some others.

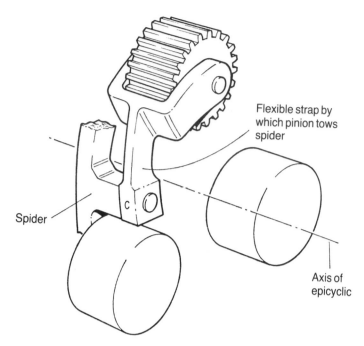

Flexible strap by
which pinion tows
spider

Spider

Axis of
epicyclic

Figure 7.6   Strap mounting of epicyclic planet.

Figure 7.6 shows a strap arrangement that could be converted into kinematic
means (2) by a bearing at C. Figure 7.7 shows a diaphragm arrangement (from the
Rolls-Royce Gazelle engine) and Figure 7.8 shows a pin arrangement.

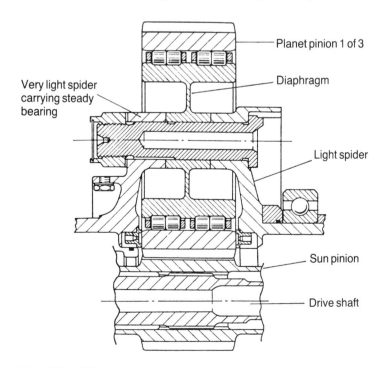

Planet pinion 1 of 3

Very light spider
carrying steady
bearing

Diaphragm

Light spider

Sun pinion

Drive shaft

Figure 7.7   'Gazelle' reduction gear. (By kind permission of Rolls-Royce Ltd.)

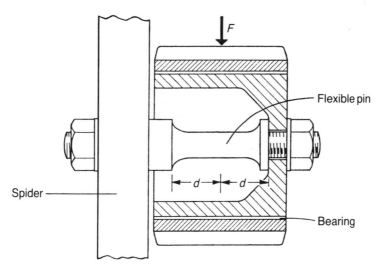

Figure 7.8   Pin mounting of epicyclic planet.

Notice that in the arrangement of Figure 7.7 the drive is fed into the mid-length of the sun pinion by a quill shaft. This is done to minimise the effects of torsional 'wind-up' in the pinion. Without the diaphragm to allow the planet to align itself, it would be better to put the torque into the end of the pinion so that the wind-up helped to cancel the effects of twist in the spider.

In compound-planet epicyclics, where the planet consists of a large-diameter gear meshing with the sun pinion secured to a small-diameter gear meshing with the annulus, a self-aligning arrangement may be impracticable. If the arrangement of Figure 7.9 is adopted, with the sun pinion train nearest to the output end, the various slopes on the teeth can be arranged so that they nearly cancel one another. For the planet wheel, the slope due to bending in the planet shaft adds to the slope due to the spider twist, and their combined effect may be largely cancelled by pinion twist. For the planet pinion, the slopes due to spider twist and shaft bending are opposite in sign and can be arranged to have a nearly zero sum (twist in the annulus is negligible).

In the simple epicyclic gear a kinematic solution is possible, though the self-aligning 'pair' may be elastic instead of kinematic. In the compound-planet epicyclic gear the solution suggested is a characteristic piece of elastic design. When neither of these kinds of solution is feasible (see next section), recourse must be had to making the teeth depart from the straight in the unloaded state, by modifying the helical angle or using longitudinal relief in addition to the usual small amount at the tooth ends. Modifying the tooth form in this way will work properly for only one load. Tooth matching is done this way in marine reduction gears where, owing to the long slender pinions, the problem is especially severe and the scope for more refined solutions is small.

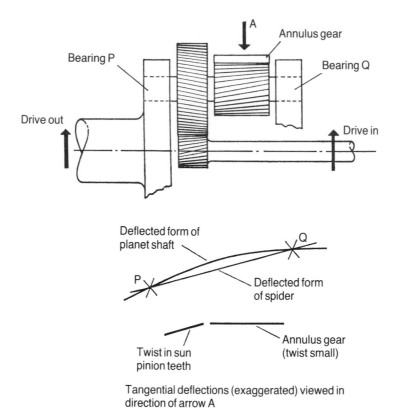

Figure 7.9   Twist in epicyclic gears.

7.3   Elastic design and elastic pairs

In the previous section it was shown how the load distribution along the teeth of a simple epicyclic can be made nearly uniform by what is virtually a kinematic design using an elastic hinge instead of a kinematic one. The load distribution along the teeth of an epicyclic with compound planets can be made nearly uniform by suitable choice of the flexibilities of the components and a folded arrangement of the trains; this second problem might be solved kinematically also, by hinging each gear of the compound planet independently and connecting them by a flexible quill shaft, but such an arrangement would almost certainly be expensive, clumsy and heavy.

Now this use of elastic components in order to ensure fairly uniform tooth loads in the compound epicyclic planet is not an approximate realisation of some kinematic design, but a fundamentally different idea of a kind we shall call 'elastic design'. It represents a diametrically opposite approach.

Kinematic design aims at a statically determinate system, where the forces in all the parts are fixed by conditions of equilibrium alone. Elastic design, on the other hand, aims at a redundant system, where the forces in the parts cannot be determined by considerations of equilibrium alone but depend upon their relative stiffnesses, i.e. equations of compatibility are involved. However, in elastic design the stiffnesses are regulated so that the compatibility conditions lead to desirable distributions of loads and forces. We might say that the principle of kinematic design is: 'Never start a fight.' On this analogy, the principle of elastic design is: 'Never start a fight unless you pick the contestants so that the outcome will be to your liking.'

One important difference between kinematic and elastic design is that small variations in dimensions do not affect load distribution in kinematic systems but they do in elastic ones; as a consequence the flexible elements of the latter must be able to absorb the effects of manufacturing tolerances without either unbalancing loads to an unacceptable extent or exceeding allowable stresses. As will be seen later, high flexibility means a large component, so that in elastic design it is sometimes necessary to choose between awkwardly large elements or rather stringent tolerances.

The self-aligning planet

Designs such as the self-aligning planet of Figure 7.7, which may be regarded as a kinematic one with an elastic element replacing a kinematic pair, occupy a half-way position. The elastic element is so much 'softer' than the spider that, to a first approximation, we can regard it as a rateless hinge, and so find the (angular) deflection in it. We can then multiply this deflection by the stiffness, and insert the moment so obtained to obtain a second approximation which is good enough for all practical purposes; a member which can be treated so is called *compliant*. In the treatment that follows the exact approach will be used.

The self-aligning planet of Figure 7.7 may be idealised as the system shown in Figure 7.10, which is a view along a radius through the centre P of the planet. The resultant of all the tooth forces could be taken as a single force, F, acting through some point on the axis of the planet (since it is an idler). It is more convenient to reduce them to a force, F, acting through P and a moment, M.

The elastic element, the diaphragm, can be regarded as a stiff hinge at P, of angular stiffness k_H, where angular stiffness is defined as restoring torque/angle turned through. The flexibility of the spider can be represented by another stiff

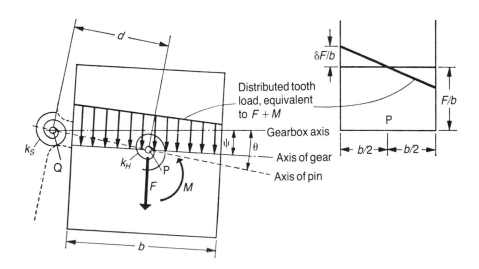

Figure 7.10   Idealisation of Figure 7.7.

hinge, Q, of angular stiffness, k_S, at a distance, d, from P; Q will be a little short of the end plate because of the bending in the pin itself.

Taking moments about P and Q,

$$\left.\begin{array}{c} M = k_H(\theta - \psi) \\ Fd - M = k_S\theta \end{array}\right\} \tag{7.1}$$

where θ and ψ are the slopes of the pin and the planet respectively, as in Figure 7.10.

For the planet to have a slope ψ, it must force the annulus and sun teeth to slope. As these are generally very stiff, ψ will be small, and for a first approximation may be taken as zero. Eliminating θ from equations 7.1 gives

$$M = \frac{Fdk_H}{k_S + k_H} \approx \frac{Fdk_H}{k_S} \tag{7.2}$$

Thus the essential requirement for good load distribution is that k_H should be small enough compared with k_S.

If the ratio of maximum to mean load per unit length is $1 + \delta$, as in Figure 7.10, the face width is b and the distribution of tooth load is linear, then it is easily shown that $\delta = 6dk_H/bk_S$, so that for a given value of δ which is regarded as acceptable, say 0.08, we can calculate the required value of k_H/k_S. We then decide how small k_H can be made (the more space there is available, the smaller this will be) and hence how large k_S must be, i.e. how stiff a spider is required.

An arbitrary decision

This seems like a rational design procedure, but the basic principle of not making arbitrary decisions (Section 9.2) was violated when the flexible diaphragm was placed in the middle plane of the planet. If it is put slightly further away from Q, so that the centre of the facewidth has a slight moment arm about it $(k_H/k_S)d$, as in Figure 7.11, then a uniform tooth loading will provide the necessary moment to align the teeth. If PQ is unaltered at d, then equations 7.1 become (with ψ = zero)

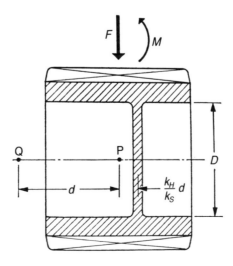

Figure 7.11   Redesigned geometry of Figure 7.7.

$$M + F\frac{k_H}{k_S}d = k_H\theta \left.\begin{array}{c} \\ \\ \end{array}\right\}$$

$$Fd - M = k_S\theta$$

(7.3)

and $M = 0$.

What has been gained by shifting the diaphragm? The facewidth of the planet can be reduced in the ratio $1 + \delta{:}1$, and this is well worth while, but the problems of designing the diaphragm and the spider are scarcely altered, even though it may seem that the requirements for k_S and k_H are less stringent. To see this, it is necessary to look at the elements themselves.

The diaphragm has to withstand a shear load, F, and deflect through an angle, θ; our improvement has not altered F, and has reduced θ only very slightly, and as these two quantities are a measure of the *task* to be performed the problem remains virtually unchanged.

Limits of elastic elements

As the diaphragm is difficult to analyse, another kind of elastic element will be used to illustrate the limitations of the tasks they can perform. Figure 7.12 shows an elastic hinge that is required to turn through an angle, θ, as large as possible, while sustaining a given force, F, along its axis. It consists of a torsion rod, AB, of fixed length, L, encastered at A, made of material with allowable stress, f, on a basis of maximum shear stress (as in Figure 3.10: Tresca criterion) and a shear modulus, G. If σ is the direct stress and τ the shear stress due to torsion, the Tresca criterion reduces to

$$\sigma^2 + 4\tau^2 = f^2$$

(7.4)

and if r is the radius of the rod,

$$\sigma = \frac{F}{\pi r^2}, \tau = \frac{G\theta r}{L}.$$

(7.5)

Substituting in equation 7.4,

$$\frac{F^2}{\pi^2 r^4} + 4\frac{G^2\theta^2 r^2}{L^2} = f^2.$$

(7.6)

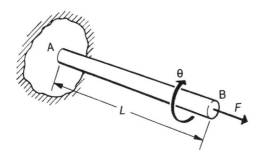

Figure 7.12   A simple elastic element.

Differentiating with respect to r, and noting that, when θ is a maximum,

$$\frac{d(\theta^2)}{dr} = 0,$$

$$\frac{-4F^2}{\pi^2 r^5} + 8\frac{G^2\theta^2 r}{L^2} = 0 \tag{7.7}$$

Eliminating r from equations 7.6 and 7.7 yields

$$\theta = 0.55\frac{f^{3/2}L}{GF^{1/2}} \tag{7.8}$$

This is the maximum value of θ that can be had under the prescribed conditions. The requirement to support a load F means that the value of θ obtainable with the given geometry and in the space available is limited.

A similar limit exists in the case of the diaphragm of Figures 7.7 and 7.11. It occupies a limited space (diameter D in Figure 7.11) and carries a shear force F, so that the value of θ is limited. From equations 7.3,

$$\theta = \frac{Fd}{k_S}$$

and this dictates the required value of spider stiffness k_S. For the central diaphragm

$$\theta = \frac{Fd}{k_S + k_H}$$

but the reduction in facewidth has reduced d, so the off-centre diaphragm is a slight help.

The result given in equation 7.8 depends on the geometry of the elastic element. If we used a spaced bundle of round rods, or a stack of flat leaves, a greater value of θ could be had for the same F and L. The task consists of a force term (F) and a displacement term (θ). If these are directly coupled, i.e. if the force is in the direction of the displacement, the element is a spring and is governed by the simple limits noted in Section 4.5. But in the more typical cases of Figure 7.13, the writer knows of no absolute limits on the task (F, θ) that can be performed by an element of given material occupying a given space. Nevertheless, it seems sure that such limits must exist, and they certainly do as far as practical designs are concerned. Unfortunately, it is often difficult to find even the practical limits applying to specific geometries—the torsion rod is not typical in this respect. Examples are given among the questions at the end of this chapter.

Sometimes the task we want the element to perform demands a low stiffness as well as or instead of a high allowable displacement. That these requirements are different can readily be seen in the case of the torsion rod. Equation 7.6 may be regarded as showing the 'Tresca strength-squared' being shared between direct and torsional loadings, in a ratio which proves to be 1:2. For a very small θ we can share the 'strength-squared' almost in the ratio 3:0, the radius can be reduced by a factor of $(1/3)^{1/4}$ and the stiffness to 1/3 of the value for the maximum θ rod. The diaphragm for the planet gear of Figure 7.7 must be designed for maximum θ, rather than low stiffness.

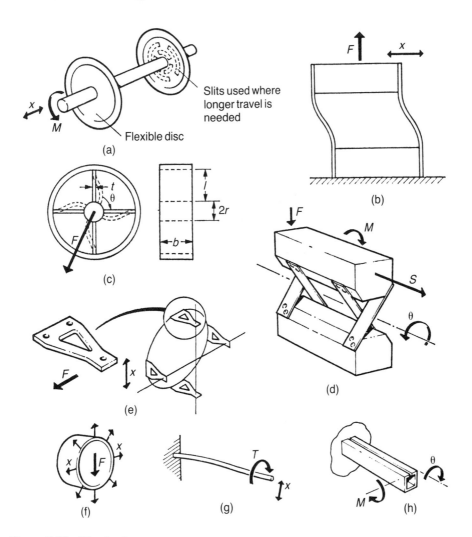

Figure 7.13   Elastic elements.

Effect of tolerances

Before leaving this example, let us look briefly at the effect that manufacturing tolerances have on load distribution. The effect of helical errors of the same hand and amount ψ (see Figure 7.10) in the annulus and sun of the epicyclic gear arranged as in Figure 7.11 is to introduce a moment, M, of roughly ψk_H. With a kinematic hinge, $k_H = 0$ and so $M = 0$, and the manufacturing tolerances have no effect on the load distribution. The tolerances also increase the task for which the elastic element must be designed.

The three arrangements for distributing the load on large tilting pad thrust bearings shown in Figure 7.2 are

(a) a pure kinematic design,
(b) a hybrid, and
(c) an elastic design.

The three kinds of load balancing possible in such arrangements are:

(1) equalising of loads between pad and pad,
(2) location of the centres of pressure on individual pads, and
(3) more detailed control of the load distribution on individual pads than is implied by (2).

Only Figure 7.2(c), with a multiplicity of supporting springs under a relatively flexible pad, can provide (3). Arrangement (a) provides (1) and (2) by purely kinematic means, while (b) provides (2) largely by kinematic means and a measure of (1) by elastic means. The springs used in (c) are heavily preloaded by central telescopic ties which collapse in compression. The task performed by these springs is measured by the force they must exert and their softness (i.e. the reciprocal of their stiffness), which is necessary to allow for manufacturing tolerances without too much variation of loading. Because the force F and the stiffness k are in the same direction, there is an absolute lower limit to the amount of material needed for a linear spring to perform this task (see Section 4.5).

Some other elastic elements

Figure 7.13 is effectively a page of the engineering repertoire, showing a selection of the commoner elastic elements. The loads shown are not the only ones the elements can sustain, but have been chosen to show possible uses.

The elastic equivalents of kinematic slides, Figure 7.13(a) and (b) are frequently used, as is the 'crossed-spring pivot' (d) widely met in instruments such as wind-tunnel balances. Figure 7.13(c) is a more expensive form of crossed-spring pivot.

One of the most useful of all elastic elements is the cantilevered spring (e)—three or four arranged as in the small inset are useful for dealing with the centring problem discussed in the next section, and two can be used in arrangement (b). The short thin walled tube (f) will serve a similar purpose, since it can be stretched radially relatively easily while being very stiff in bending and shear.

The elements shown in Figure 7.13(g) and (h) are complementary in function—the first transmits a torque while permitting small transverse movements, while the second is easily twisted but stiff in bending.

7.4   Comparison of kinematic and elastic means

There are a few other, rare, means of guiding or locating one body relative to another, such as electromagnetic induction. We will not include hydrodynamic and hydrostatic bearings, which are effectively kinematic, so short is the distance between the mating surfaces. In the majority of cases the only practical choice is a kinematic means, but there are many interesting problems where both kinematic and elastic means are worth study.

Advantages of kinematic devices

(1) Smaller when 'displacement term' is not very small.
(2) Free of stiffness (zero rate) in desired direction of movement.
(3) Available as standard items.

(4) High stiffness in 'fixed' directions.
(5) Load-balancing functions not affected by tolerances.

Disadvantages of kinematic devices

(1) Larger when displacement is very small.
(2) Subject to hysteresis *or* dependent upon pressurised lubrication *or* requiring to be kept moving *or* must be of rolling type.
(3) Subject to backlash.
(4) Difficult to protect from corrosion and dirt, and sometimes 'brinelling'.
(5) Requiring maintenance.

Advantages of elastic devices

(1) Smaller when displacement is very small and a high ratio of stiffness in 'fixed' and displacement directions is not required.
(2) Free of hysteresis and backlash.
(3) Requiring no maintenance.
(4) 'Cut-price' solutions are sometimes possible.

Disadvantages of elastic devices

(1) Larger when displacement is large or stiffness must be low, or ratio of stiffnesses must be large.
(2) Stiff in displacement direction.
(3) Sometimes expensive.
(4) Careful and expensive design sometimes needed to avoid fretting.
(5) Load balancing upset by tolerances.

A few cases will be discussed.

Further discussion of load sharing in gears

It is economical in gearing with high maximum torques to use multiple layshafts, and the planets and compound planets of epicyclic gears are, in effect, multiple layshafts. It has already been shown that load sharing between the layshafts in simple epicyclics can be achieved kinematically by floating sun or annulus, or elastically by flexibility in the annulus. With more than three planets, the elastic method still works but the kinematic does not (Section 8.2).

With multiple layshafts that are not epicyclic in action, a common number is two. Load sharing here is generally done by flexible quill-shafts between layshaft pinions and wheels, but kinematic means have been used, e.g. by end thrust balancing with helical gearing. Notice that the flexible layshafts represent a true elastic design, not an approximation to a kinematic design.

Centring devices

The bores of turbine and compressor discs may grow relative to their shafts due to centrifugal and thermal stresses by as much as 0.003 times their diameter.

Figure 7.14(a) to (d) shows ways of holding the discs concentric with the shafts

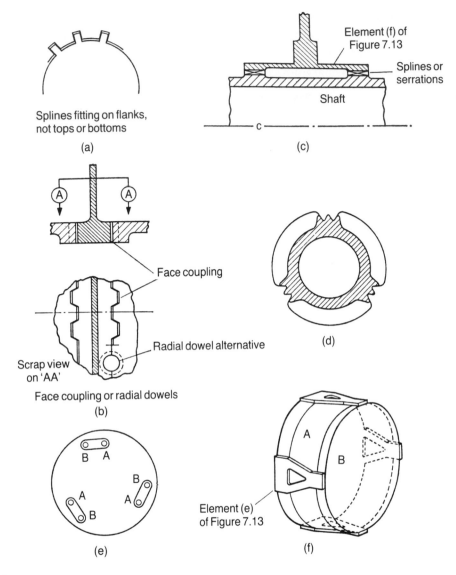

Figure 7.14   Centring devices.

under these circumstances. Diagrams (a) and (b) show kinematic means, which require very high standards of accuracy in manufacture; the face-serrations (b) are ill-adapted to the centring of long stacks of discs on a separate shaft. Cheap versions of (b) may be had by inserting radial dowels with their axes lying in the plane of two butted faces.

The simplest elastic solution is a large interference. Where this is not acceptable, the arrangement shown in Figure 7.14(c) may be possible. A similar solution is shown in Figure 2.4. The serrated ribs in (d), an unproven solution systematically invented by students under the writer's guidance, provide local centrifugal loads which bend the shaft wall so that the small initial interference is not lost. The writer believes that this may be the best solution where the shaft wall thickness is small, but where the disposition problem of Section 6.3 arises it is difficult to find an embodiment which is both light and simple to make. Because the shaft wall is bent, and not stretched, ribs of quite small section suffice.

The same centring problem arises with stationary components owing to differential expansions. Freed of the high centrifugal field, a wider range of solutions is possible, including those shown in Figure 7.14(e) and (f), in both of which A and B denote the two parts to be kept concentric. In the kinematic solution (e), the pins A and B are in parts A and B, respectively. Eccentrics may be used instead of links, giving a compact embodiment, provided it is sure they will overhaul. The elastic solution (f) uses four elements (three would suffice) of the simple cantilever spring sort shown in Figure 7.13(e).

Instruments

In many instruments, hysteresis and backlash in pivots and slides are much more objectionable than a slight rate, and circumstances often strongly favour the use of elastic elements. Often both the displacement and the load term in the task are small, and the stiffness can also be made small; sometimes, as in galvanometers, the stiffness can be made use of (or a cross-stiffness, as in some twisted strip comparators which also use an elastic slide).

Hinges in gas-cooled reactor ducts

In the early gas-cooled reactors for power generation, where welded steel pressure vessels are used, thermal-expansion difficulties were overcome by hinging the ducts, using bellows to provide sealing. The hinges themselves were either simple pin hinges, outside the ducts, flexible strips inside the ducts, or elastic hinges outside the ducts, the last of very elegant design. A full account of this interesting design problem and the various solutions adopted is given in reference 51.

7.5   Elastic design and structural economy

The principal advantage of kinematic design is its determinacy, the independence of the loads in the parts of elastic considerations. This advantage is only to be had at a considerable structural cost, i.e. much more material is needed than for a non-kinematic or elastic design. Sometimes this cost will not be significant in financial terms, but in all very large things it will. Even in quite small objects, the structural economy of elastic design may be of crucial importance, as the following example will show.

A roller bearing is an elastic design; it is statically redundant, and the distribution of the load between the rollers depends on their elastic deflection. We could make a kinematic roller bearing with three rollers; it would be about seven times the volume and would have several other very undesirable attributes. Notice in passing that we can improve the load sharing (or increase the tolerances on the components) in a roller bearing if the elasticity is increased, e.g. by making a hole through the middle of the elements. This has been done both in parallel and taper roller bearings. Notice also that the lengthwise distribution of load on a cylindrical roller on a cylindrical race will have peaks at the ends because of variation of the stiffness due to three-dimensional effects. This can be corrected by the last of the three means given in Section 7.2, by making the rollers not truly cylindrical but slightly relieved or 'blended' at the ends.

The helicopter engine in Figure 7.1 is a massive object traversing a clear space in the fuselage; it seems a waste that it cannot be used as a structural component of the airframe. Apart from the extra difficulty set by the large thermal expansion of this potential member, this is a degree of sophistication not yet reached in aircraft design.

Structural economy

Consider a very simple structure, a uniform horizontal beam, AB, subject to a single vertical load W. Figure 7.15 shows two versions, (*upper*) pinned at both ends, i.e. kinematic (except for the redundant constraint in the AB direction, which need not concern us) and (*lower*) a redundant version encastered at both ends. Wherever W acts, the bending moment is less in the encastered beam because it is receiving more help from its supports, which in a mechanical-engineering application will generally be further parts of the structure. Even a pin-jointed frame with given nodes, if subject to different systems of loads at different times, will probably take less material if it is redundant [17]. Suppose such a frame is fixed at A and B, and loads W_1 and W_2 are to be applied at C and D (Figure 7.16). Then if a frame is provided to support W_1 at C from A and B, part of the load W_2 can be routed via C. For the particular configuration shown, it looks very likely that it will be more economical to route as much possible via C. Depending on the ratio of W_1 and W_2, it may not be possible to take all the load via C, and then rather than increase the scantlings of the members AC and BC say, it may be cheaper to go straight to A. We now have a redundant structure, and some of W_1 will go by D, so that AC and BC should perhaps be thinned—the problem is very complicated, but the key point is this: the route to 'earth' from D is cheaper via C *only* when part of that route, from C to earth, is an existing *asset* (see Section 8.1).

Note also that in the frame of Figure 7.16 it might pay to have extra nodes between CD and AB, depending on end costs and the structure loading coefficient. The structure loading coefficient, already mentioned in Section 6.7, is a parameter expressing the size of the loads relative to the size of the structure and the strength of its material. In this case we might use the dimensionless form

$$\frac{W_1}{(AB)^2 f}$$

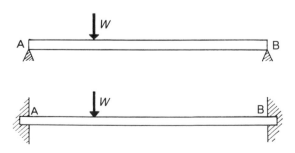

Figure 7.15   Redundancy and structural economy.

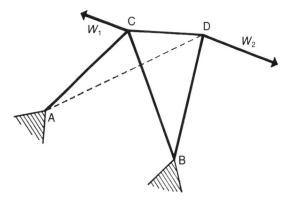

Figure 7.16   Structure economy with different load systems.

where f is the allowable stress. If the structure loading coefficient is low, the scantlings are small compared with AB, the members are slender and so buckling is an important consideration, favouring extra nodes in this case, and more elaborate structures in general.

Load diffusion

Another way of seeing the structural superiority of elastic design is via the idea of 'load diffusion'. If we wish to locate one body, X, relative to another, Y, by a kinematic design, all the loads in X must be gathered up into a relatively few relatively small zones which are attached to Y, and then diffused out again all over Y. If we look outwards, as it were, from the attachment points, we have two load diffusion problems. The cost of load diffusion will be high when the structural loading coefficient is low. A good example is balloon design, where the buoyant forces spread over a vast area of flimsy envelope must be gathered up to support a small dense load. A parallel is the supporting of the spherical propellent tank in the ELDO rocket third stage by 192 titanium alloy straps—notice the structure loading coefficient here was low largely because of a very high allowable stress.

Where the structure loading coefficient is high, the cost of gathering up loads is much less; however, once gathered it may be worth keeping them so, as in those diesel engines that use stiff-jointed frames for their structure [33]. Non-structural casings close in the openings in these frames, a case of separation of function.

With the tendency of engineering devices to become bigger and to use smaller quantities of stronger but more expensive materials, the relative number of cases is increasing where the structural economy of elastic design is more valuable than the determinacy of kinematic design.

Questions

Q.7.1(2). Figure 7.17 shows a section commonly used for the slideways of lathes (only the saddle ways are shown). Alignment tests on the accuracy of lathes call for checks to be made of the levelness of the front (inverted V) and back (flat) ways

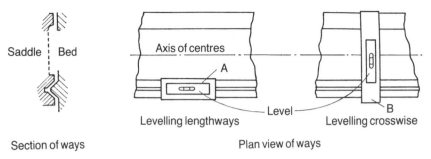

Saddle | Bed

Axis of centres

A

Level

B

Levelling lengthways

Levelling crosswise

Section of ways

Plan view of ways

Figure 7.17   Alignment tests on lathe.

separately, and also of the truth of front to back ways by means of a bridge piece straddling them, a sort of dummy saddle. Think about this with regard to (a) the value of such tests, (b) the importance of the different kinds of error which might be found and (c) the nature of the pieces A and B (Figure 7.17) used to adapt the sensitive level to these measurements.

Q.7.2(2). Figure 7.18 shows a duct subject to an internal gauge pressure p, at a point where an elastic hinge is provided. The continuity of the pressure wall is maintained by a bellows which is relieved of the axial pressure load $p\pi R^2$ by a flexible tongue of effective length L anchored in two spiders S_1, S_2. This tongue and the bellows permit flexure to an angle α from the straight, necessary to accommodate differential thermal expansions in the system of which it forms a part. Estimate very roughly the necessary length L, assuming the hinge is free from shear forces, or axial forces in excess of that due to internal pressure. The allowable stress in the material of the tongue is f and the Young's modulus is E.

Q.7.3(2). Consider the design of the flexible pin in the self-aligning pinion arrangement of Figure 7.8, in particular the optimum variation of section along its length. Do not plunge into a flurry of mathematics, but see first how far common sense will take you. For example, is there any portion of the length in which an infinite EI would be desirable? Determine a practical optimum form, bearing in mind the stubbiness of the pin, and then derive an approximate attainable θ in terms of the notation of Section 7.3, assuming the effective length of the pin is $2d$ symmetrically disposed about the mid-length of the planet pinion.

Q.7.4(2). Figure 7.19 shows a three-bearing shaft assembly, with bearings at A, B,

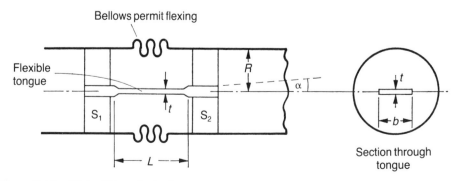

Bellows permit flexing

Flexible
tongue

S_1

t

S_2

R

α

t

b

Section through
tongue

L

Figure 7.18   Virtual hinge in duct.

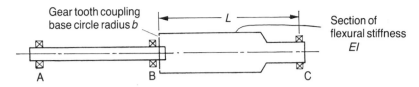

Figure 7.19   Three-bearing shaft.

and C. Close to B there is a gear tooth coupling of base circle radius b, with a coefficient of friction μ between the teeth. The portion AB of the shaft may be regarded as stiff, and the portion BC, of length L, has a flexural stiffness EI. The torque in the shaft at the gear tooth coupling is T. Estimate (a) the axial force the gear tooth coupling can sustain without slipping and (b) the amount, δ, that bearing C must lie off the axis of AB for the misalignment to cause slipping in the gear-tooth coupling (now assumed free of axial load).

Q.7.5(3). Consider the problem of ensuring equal sharing of the loads in a simple epicyclic with n planets, where $n > 3$. How will the efficacy of a flexible annulus gear vary with n? What other means are available?

Q.7.6(1). A twin layshaft reduction gear of the layout shown in Figure 7.20 is such that the sum of the torques in the two layshafts is $2T$; if all the elements concerned are at the appropriate extremes of their tolerances, and a light torque is applied to the input shaft, the lagging layshaft pinion has to be advanced an angle ϵ to make contact on the driving flanks.

Using the allowable stress diagram of Figure 3.10, find the shortest distance L with which it is possible to ensure that the torque distribution is not more unbalanced than $1.1T$ to $0.9T$.

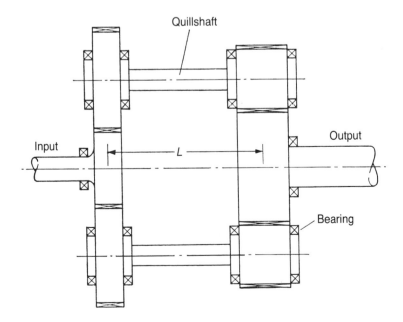

Figure 7.20   Twin layshaft reduction gear.

Answers

A.7.1(a). The writer doubts the value of such tests, since all that is of practical importance is the truth of the saddle motion, which should be measured directly. It can be argued, however, that a fortuitously or deliberately produced arrangement of high spots could give a true saddle motion on irregular ways; the high spots would soon wear, and the quality of the lathe would rapidly deteriorate. In effect, this is an argument that the tests are necessary to check the quality of the guiding surfaces, in which case it would surely be better to use tests specifically designed for the purpose.

(b) If the front and back ways rise and fall together, the tool merely rises and falls and there is no practical effect on the accuracy of the work. If, however, the ways cross-wind or warp, the front rising when the back falls and vice versa, the saddle will roll and the tool move in and out of the work.

(c) The pieces A and B should both be constrained to have only one degree of freedom, i.e. translation along the ways. Accordingly five of their potential six degrees of freedom must be suppressed by a total of five simple constraints. For piece A these can be provided by three balls F (Figure 7.21) and a roller R, the former providing three simple constraints, and the latter two simple constraints, since a line is constrained to lie in a plane.

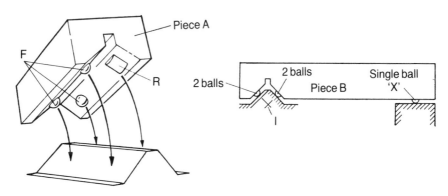

Figure 7.21   Alignment tests on lathe.

Piece B is a little more difficult. At least one constraint must be associated with the back way, otherwise the motion will be independent of it. There cannot be as many as three associated with the back way, since it is a nominal plane and with three constraints would limit the motions of B to two translations and rotation about a vertical axis; the important displacement to detect, however, is rotation about an axis parallel to the headstock, as has been discussed under (b). There must be one or two simple constraints on the back way.

With one simple constraint at the back way, we need four at the front, and this can be achieved by substituting another ball for the roller in piece A (Figure 7.21). Piece B is then free to rotate about the instantaneous axis, I, associated with the four balls on the front way until ball X rests on the back way.

With two simple constraints on the back way, say two spaced balls at X, then only three balls must be used on the front way, and an additional closing force must be used to keep the third ball in contact with the bed, just as a T-square must be kept pressed against the edge of the drawing board. Thus the former arrangement, with only one ball on the back way, is to be preferred.

A.7.2.

$$L = \frac{2\pi p\,R\alpha E}{kf^2}, \text{where } b = kR$$

It is clearly desirable to make b as large as possible, and the limit is $2R$. It is therefore appropriate to express b as a multiple of R, kR, where $k < 2$.

As the highest bending stress associated with α and the direct stress add at some points, their sum must be equal to f. The curvature is α/L so that the maximum bending stress is $\alpha Et/2L$.

$$\text{Then } f = \frac{p\pi R^2}{kRt} + \frac{\alpha Et}{2L} \tag{7.9}$$

Since f is fixed, equation 7.9 can be differentiated with respect to t, giving

$$0 = -\frac{p\pi R}{kt^2} + \frac{\alpha E}{2L} - \frac{\alpha Et}{2L^2}\frac{dL}{dt} \tag{7.10}$$

and when L is a minimum $dL/dt = 0$. Either by this route, or simply by recognising the equal moduli of the powers of t in equation 7.9, it is found that L is a minimum when f is equally shared between bending and direct stress (see Section 6.10) and hence the required answer can be obtained.

In practice, it would be necessary to allow for shear loads, S, on the hinge, and also for secondary bending moments due to the deflection. The bending stress due to the shear load increases with L, thus imposing a limit on what can be done in the way of reducing stresses by making the tongue longer, and the task represented by $(p\pi R^2, S, \alpha)$ may be beyond any available material. It would then be necessary to adopt some other solution [51].

Notice that in this problem a figure of merit for the material is f^2/E, which is also a figure of merit for a spring material (Section 4.5). Note also that by 'building bent' it might in some cases be contrived that the bending stresses alone were present when cold and depressurised and the direct stresses alone when hot, or some variation on this theme.

A.7.3. Since only curvature which produces a slope opposed to θ is helpful, the portion of the pin from the line of action of F to the spider should be as stiff as possible. The remainder should be fully stressed, which means that it should be thinnest on the line of action of F, and increased in diameter to the right (Figure 7.8), in order to withstand the increasing bending moment. Allowing for the stubbiness of the pin, however, it is doubtful whether it is practicable to exploit this 'ideal' form, and a good enough approximation is probably achieved with two diameters, as large as possible to the left of F and as small as possible to the right.

The moment at the free end of the pin is Fd, so that the radius, r, of the thin part of the pin is given by

$$f = Fd/\tfrac{1}{4}\pi r^3 \tag{7.11}$$

The slope developed in the length d of the thin part is

$$\frac{Fd^2}{2EI} = \frac{2Fd^2}{\pi Er^4} \tag{7.12}$$

Putting this slope equal to θ and using equation 7.11 gives

$$\theta = \frac{df}{2E}\left(\frac{\pi f}{4Fd}\right)^{1/3}$$

A.7.4. (a) $\mu T/b$. (b) Of the order

$$\frac{2\mu TpL^2}{3\pi bEI}$$

where p is the pitch circle radius.

The loads on the teeth total T/b, so the force required to slide them axially is $\mu T/b$. If the total tooth load T/b is distributed uniformly round the gear tooth coupling, then to 'bend' it about the horizontal diameter, say, requires all the teeth in the top half to slide one way and all the teeth in the bottom half the other. If p is the pitch circle radius, the moment arm of each set of forces about the diameter is $(2/\pi)p$, so that the resisting moment is

$$2 \times \frac{\mu T}{2b} \times \frac{2}{\pi}p \ (\text{see Figure 7.22})$$

If the shaft BC is regarded as encastered at B and given a deflection δ at C by a load F,

$$\delta = \frac{FL^3}{3EI}$$

and the moment at B is

$$FL = \frac{2\mu Tp}{\pi b}$$

Once the coupling slips, however, and takes on an angle, the skewing of the teeth relative to one another will concentrate the load in two arcs near the ends of the horizontal diameter, greatly reducing the resisting torque.

These rough calculations, with values substituted, will indicate whether it is safe to rely on a gear tooth coupling or serration to allow axial sliding. The writer made an experiment on a running gas turbine with the configuration of Figure 7.19 and

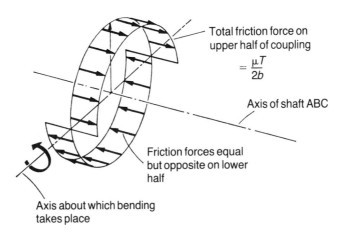

Total friction force on upper half of coupling

$$= \frac{\mu T}{2b}$$

Axis of shaft ABC

Friction forces equal but opposite on lower half

Axis about which bending takes place

Figure 7.22   Bending of a coupling.

found that the force needed to slide the coupling corresponded to three times the capacity of the separate thrust bearings on the shafts AB and BC; this would agree with the expression $\mu T/b$ with a high value of μ (about 0.5). The hope that gradual slip would occur because of wriggling of the teeth by misalignment of the bearings proved ill-founded, as might have been shown by the sort of calculation given above. The δ required was larger than would normally occur.

A.7.5. A full analysis of this problem will not be given, but here are a few notes. The load-sharing properties of a flexible annulus gear depend on the tooth load at any mesh causing a deflection which reduces the 'interference' as it were, at that mesh. There are two such flexibilities, one associated with the radial component and the other with the tangential component of tooth load. The effect of the former may be likened to that of an elastic band round a nominally symmetrical bundle of pencils, where even if the pencils vary slightly in diameter the force exerted on them by the band is nearly the same (notice, however, the band works in tension and the gear in bending).

A study will show that the stiffness to the radial component approaches proportionality to n^2: while it may well be low enough at $n = 4$ to effect load sharing, at $n = 8$ it will be about four times as great and so much less useful.

The other flexibility arises from the moment arm of the tangential component of tooth load about the principal axis of the ring cross-section; the ring bends into a wavy form with n wavelengths in the circumference. This form of flexibility decreases only as n^{-1}, but is smaller in the first place, e.g. at $n = 4$.

Thus it is clear that as n increases, the flexibility of the annulus decreases in efficacy as a load-sharing agency. The point at which it becomes inadequate will depend on the accuracy of the gears, the level of stressing, and the fineness of the teeth.

When annulus gear flexibility is not enough, springs between the separate planets and the spider can be used, e.g. the arrangement of Figure 7.8 will serve this purpose as well as aligning the planets, and in this respect it is superior to that of Figure 7.7. Alternatively, hydraulic means between the planets and the spider might conceivably be used, as might kinematic means.

A.7.6

12.3 $\dfrac{GT^{1/3}\epsilon}{f^{4/3}}$ where G is the shear modulus.

The difference between $1.1T$ and $0.9T$ must produce a wind-up ϵ in a quill shaft. The quill shaft must also be able to withstand $1.1T$. To be as flexible as possible for a given strength it must be solid. If the quill shaft radius is r, then

$$\epsilon = \frac{0.2TL}{\frac{1}{2}\pi r^4 G} \cdot \frac{f}{2} = \frac{1.1T}{\frac{1}{2}\pi r^3}$$

Eliminating r gives the answer. Notice that the figure of merit here is $f^{4/3}/G$, as in the elastic element of Question 7.3 except for a different modulus. Note also that the form $f^{3/2}/G$ arises in the study of the element of Section 7.3 (equation 7.8) and f^2/E in the element of Question 7.2.

8 Costs

8.1 Design and costs: tasks, values and assets

There is a great deal of truth in the aphorism, 'An engineer is a man who can do for one dollar what any fool can do for two', and no designer should be ashamed that the reduction of cost is the chief test of his skill. However, the cost in question must be the true or total costs for the same benefit, not just the purchase price, and must include running costs, maintenance, the costs of unreliability and so on. The means of making such costs commensurate are not part of design, but one important aspect has been covered in Section 3.2. The benefit is also difficult to assess in some cases, particularly in consumer goods, where it is hard to judge such things as the money value to the customer of a sunshine roof in a car.

Although all kinds of engineer are involved in keeping costs down and value up, it remains true that once the design leaves the design office it is difficult thereafter to make substantial reductions in cost. Moreover, it is in the early stages of design that the final cost is largely settled. What are required are reliable ways of comparing costs of alternative schemes in the early stages, and these are not yet widely available. One difficulty is the complicated interaction involved where the cost advantage of a different concept may lie in making a different method of manufacture possible for some component. Nevertheless, in some rather specialised areas using rather standardised production methods, computer programs are used with great success to predict costs from a very early stage in design.

Besides keeping the costs continually under review throughout design, which is a counsel of perfection, it is common to examine the costs at one point, preferably in the embodiment stage, and to try to reduce them, usually by changes in the form design, the method of production or the material. Value engineering is such an approach, and it has proved its usefulness. However, the whole subject of costing in design is relatively little developed, considering its great commercial importance.

One central problem is that costs are largely fixed at the conceptual stage, while costing by conventional means only becomes possible late in the embodiment, and is sometimes left until detailing has been done. What is needed are reliable techniques for costing much earlier on in the design process.

One idea which may become important is that of costing functions, so that a first cost estimate can be made from an analysis of function, at the very beginning of the conceptual stage. Such an estimate would normally be a lower bound on the eventual cost, since the cost attributed to each function would be the lowest known to be possible, and so might not prove to be attainable in the particular circumstances of the design.

It is a help in costing a function if it can be quantified, and when this is done it will sometimes happen that the cost will be roughly proportional to the quantity.

Tasks

Several instances have been given in this book of the quantification of a function to be performed; the term 'task' is used to denote a quantified function. Section 1.9 introduced the idea of the quantification of the resistant function of a pressure vessel as holding-together power, and in Section 2.5 the possibility of quantifying the structural functions in an LNG tanker was mentioned. Neither in the case of the tanker nor that of the beam mentioned in Section 4.6 was the quantification carried out—it is fairly easy for the former and difficult and uncertain for the latter [17]. In Section 4.5 the liquefaction of nitrogen was regarded as the task of increasing its availability function, b, by a certain amount. In Section 7.3 the task of an elastic element was stated in terms of a force F and displacement θ, but the two were not combined into a single quantity. Once we can quantify functions, we can often estimate the costs of performing them. For example, for steel structures erected on site (shipyards produce more cheaply) we may for £1 get up to 28 kilojoules (0.24×10^6 lbf in.) of holding-together power if we can work at the favourable point A of Figure 3.10, or 14 kJ (0.12×10^6 lbf in.) at the less favourable but much commoner working point B.

In most practical structures we cannot even hope to work all the material at B, and an efficiency of use will have to be guessed (see Section 4.5). Nevertheless, we can estimate likely task:cost ratios closely enough for many of the purposes of conceptual design.

Values

The ability to perform a task we call a value, but it is convenient to use the term rather widely. For instance, in disposition problems we saw how different portions of a space to be disposed of had different values for different purposes—the material they could accommodate would be able to perform different amounts of the various tasks in hand. In the simple problem of the splined shaft with turbine discs (Section 6.3), bits of space in the annular region r_1 to r_0 occupied by the splines had great value for shaft stiffening but none for the disc-holding-together task.

Assets

An asset is an existing value provided for some other purpose. Thus, in the frame sketched in Figure 7.16, the structure used to support the load at W_1 at C is an asset when we come to design structure to support a load W_2 at D (W_1 and W_2 act at different times).

The asset outlook is specially valuable when a scheme has reluctantly been increased in complexity. It is natural for the designer to resist additions to his first concept that are forced upon him by more detailed study. But once it has been accepted that such an addition is essential, he should then try to draw what further advantage he can from it. In comparing alternative schemes it is easy to remark that one involves some extra component, but not so easy to make sure that that component is properly regarded as an asset and exploited to the full. Also, some devices have inherent in them assets which cannot ordinarily be turned to full account, but may be exploited fully in special circumstances (see Section 8.3).

8.2   Tasks and values in gearing

A good simple measure of the task performed by reduction gearing is the output torque. In Figure 8.1 the cost per unit of output torque is plotted against reduction ratio for data from two sources [34]. The near constancy of the cost per unit of torque shows that it is an apt measure of the task.

To see why this should be so, refer back to the scaling study of gears in Section 4.7, where it was shown that the allowable torque T_p in a pinion was given roughly by

$$T_p = \frac{CD_p^2 L}{1+k} \tag{8.1}$$

where C is a constant, D_p the pinion pitch diameter, L the facewidth and k the ratio of the number of teeth in the pinion to the number of teeth in the gear ($k < 1$).

If T_0 is the output torque, then $T_p = kT_0$. Equation 8.1 gives

$$D_p^2 L = \frac{(1+k)kT_0}{C}$$

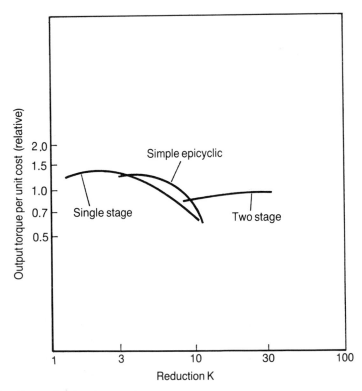

Figure 8.1   Costs of reduction gearing.

and the volume of the pinion is a fair measure of its cost. As the teeth of the output gear are similar to those of the pinion, it is reasonable to assume that its cost is about $1/k$ times as much, and so we may expect the cost of the output train to be roughly proportional to

$$\left(1 + \frac{1}{k}\right)(1 + k)kT_0 = (1 + k)^2 T_0 \tag{8.2}$$

Not much weight is to be placed on this particular result—there should be several terms allowing for different parts of the cost—but the proportionality to T_0 and the slow variation with k should be noted. When the other terms are added, there is a reversal of the steady reduction of expression 8.2 at small enough ks. Small ks will mean very large diameter gears that cannot be costed simply per yard of rim, as it were, and the cases to hold them will also increase rapidly in size [35, 50]. The effective cost of the train must start to rise at some value of k, and it must also become very high for very small ks, so that it will pay to use more stages of reduction. A first step in improving expression 8.2 is to multiply it by a factor $(1 + a/k)$, where the a/k term allows for items such as case ends and wheel webs whose cost is proportional to the area of the gears viewed axially rather than their perimeter.

We can expect a minimum cost for a train of given output torque at some value of k, k_M say. When the overall reduction ratio K ($K < 1$) is sufficiently small to require two stages, a simple optimisation will divide the ratio between them in such a way that the output stage has a ratio closer to k_M than that of the input stage, since it is larger and more expensive and a fractional increase in its cost is more significant. However, there is a large saving in making the centre distance on both stages the same, so that the bores for the input and output shafts are concentric. With this extra constraint and an empirical value for a, the balance of facewidths is determined. Lower values of k must be expected to give a somewhat higher cost per unit of output torque because they lead, first to uneconomical values of k in the stages, and eventually to the adoption of an extra stage with a net increase of cost despite the return of economical ks.

Reducing the cost of gearing

The cost of reduction gearing is closely associated with the output torque, largely because of the need for a massive final-stage wheel to carry that torque. The most promising way of reducing cost would seem to be to reduce the size and cost of this 'bull wheel'.

In order to do this we might:

(1) use the potential value of the teeth more fully, by better load distribution both along the length and through the mesh,
(2) increase the value of the teeth by changing their form,
(3) increase the number of loads on the bull wheel.

Let us examine these possibilities in turn.
(1) We have seen how to achieve better load distribution along the teeth in Section 7.2, and reference 36 gives an impressive account of improvements effected in naval gears in this way. Design for the development of the full value by controlling the variation of the load as the teeth pass through the mesh is described in reference 20.

(2) Section 6.9 discusses the problem of increasing the value of involute spur gear teeth, and much of this will read across into helical gearing. We might also consider a generalised tooth form, not necessarily involute, and maximise its strengths, a procedure which leads inexorably to short-path and point-contact gears, of which the Wildhaber-Novikov forms are typical [29]. Such tooth forms have a large inherent advantage in strength which can be exploited in at least two ways.

We can either make the gears much smaller, or make them of much weaker and cheaper stuff, or we can follow a middle course of smaller gears of somewhat weaker stuff. Here it is important to note that the load capacity of a given tooth contact is roughly proportional to the square of the material strength (Section 4.8), so that as we use cheaper stuff, surface strength is reduced more than bending strength, and it is the latter which is more difficult to provide in circular arc tooth forms.

(3) If the number of loads on the bull wheel can be increased, its value increases almost in proportion. If we provide two pinions driving the bull wheel, we can reduce the facewidth to about 55 per cent—the odd 5 per cent being needed because the required life in cycles is doubled. If only bending strength is concerned, we can win back 2 per cent by rebalancing strengths between wheel and pinion as explained in Section 6.9, but this is not possible with surface strength. An allowance for imperfect balance of loads between the pinions should be included— this is very large in marine reduction gears (the writer suspects the allowance is excessive and a vestige of ancient prejudices, especially since the longitudinal distribution of load will normally be improved).

Just to reduce the facewidth would be to squander the enhanced value; in accordance with the principle of Section 3.3, that any change must be followed through in all its consequences, the reduction should be shared between the facewidth, the diameters and probably the tooth pitch.

Notice that in such a change the facewidth:diameter ratio might be expected to be little altered, or perhaps slightly reduced. If it was optimum before, it should be optimum after, from scaling considerations (Section 4.7), but since casing size will now have a reduced effect there may be a slight shift towards a relatively narrower gear. Thus of the 45 per cent reduction rather less than two-thirds may go to decreasing the square of the diameter and rather more than one-third on making the facewidth smaller, giving perhaps 0.83 of the diameter with one mesh and 0.80 of the facewidth ($0.83^2 \times 0.80 \simeq 0.55$). Some advantages may be expected due to improvement of the longtitudinal distribution of load.

Using twin layshafts, multiple layshafts, epicyclics or star gears will reduce the gross amount of gear, but it will add to the prime cost due to all the extra components involved. Accordingly multiple meshing will only pay under certain favourable conditions, such as:

(a) very large tasks (output torques),
(b) a premium on low weight or size (aircraft, mining machinery, rail and road transmissions),
(c) large-scale production, which often favours the use of numerous smaller components.

A point which should be made about epicyclic gears is that the value of the annulus gear (its task-performing capacity or torque capacity) is less than the output torque T_0 by the amount of the input torque. This is not the case with star gears, which do not, however, suffer from centrifugal loading of the planet bearing, which is severe in such designs as that shown in Figure 7.7; in that particular case, centrifugal loads add to the stringency of the design problem set by the elastic diaphragm.

An extreme case of multiple meshing was the simple epicyclic reduction gear of the Wright 'Turbo-Cyclone' turbo-supercharged piston aero-engine, with 18 planet pinions.

8.3 Gears for contra-rotating propellers

It was mentioned in Section 5.4 that contra-rotating propellers for ships showed advantages in propulsive efficiency. Let us consider the likely cost of gears for such propellers in the light of the last section.

It will be assumed that the propeller speeds are the same as for the original single screw, and that the shaft power is unaltered. Then if T_0 was the output torque for the original gearing, the new gearing puts out equal and opposite torques of $\frac{1}{2}T_0$. Are we to assess the task at $\frac{1}{2}T_0$, or $2 \times \frac{1}{2}T_0$?

The answer to this question depends on how we generate our two 'half torques'. If it is done in separate bull wheels separately driven, the answer is $2 \times \frac{1}{2}T_0$, but if we make one 'half torque' the reaction to the other, the answer is $\frac{1}{2}T_0$.

We have only two viable means of producing these torques in the propeller shafts—by loads on gear teeth attached to them, or by loads on bearings of planetary shafts attached to them. In the simple epicyclic or star gear we have both—the annulus-bull wheel, and a set of orbiting or potentially orbiting shafts. One of these torque capacities—the 'spider' of the star gear or the annulus of the epicyclic—has a large untapped value, an asset we can now exploit. All we need do is attach the spider to the inner shaft and the annulus to the outer shaft as in Figure 8.2. It is true that the torques will differ by an amount equal to the input torque, and

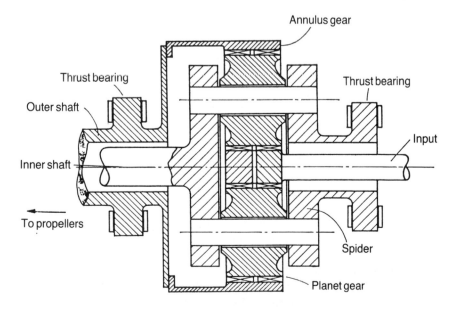

Figure 8.2   Reduction gear for contraprops.

the speeds are not precisely fixed nor the propeller phases related, so that the blade-past-hull and blade-past-blade frequencies are indeterminate—probably none of these effects is serious. If we were worried about them they could be rectified, at a cost.

The task of the contraprop gearing in an arrangement like Figure 8.2 is only about $\frac{1}{2}T_0$, so that on straight gear cost we should expect a substantial decrease. The costs of shafts and thrust bearings will be higher than for the single screw, and some of the difficulties of assembly etc. will be reflected in the casing and spider costs. Above all, the structural-cum-topological difficulties of the star gear are bound to be expensive; in particular, one of these difficulties is easily overlooked, and it is worth digressing a little to examine it.

It is practically essential to have some bearing between the inner and outer shafts. Unfortunately, plain bearings between contra-rotating shafts do not work when, as in this case, the principal load does not rotate (here it is the deadweight of the shaft and propeller). This may be seen by giving the whole system a rotation so as to make the outer shaft stationary. The load is then rotating at half the speed of the inner shaft, and will force a half-speed whirl. (If we take the Reynold's equation as a starting point, it is readily seen that neither squeeze nor sliding wedge action is present.) One way round this problem is an intermediate stationary shell, but with the arrangement of Figure 8.2 it is awkward to prevent it rotating.

To guard against overlooking this kind of problem is very difficult; check lists offer some protection against the more common errors, like designs that cannot be assembled, but a list which catered for all contingencies would be unmanageable. One of the best safeguards is to try and visualise the working of the design in all its details—to 'think through' all its aspects, however trivial they may seem.

8.4   Large pressure vessels for gas-cooled reactors

As has been shown in several places in this book, costs tend to consist of several terms varying in different ways, often roughly expressible as simple powers of the parameters defining the particular design situation. As the parameters change, the relative size of these various costs may change, though often the effect of optimisation procedures is to keep the ratios between them constant. However, as between two different solutions of a particular type of design problem, variation of the defining parameters may cause very different changes in the total costs, so that the cheaper becomes the dearer. This has already been seen in the rather special case of the aircraft compressor rotor in Section 2.3 where cost could practically be equated to weight. In this section another simple case will be examined.

The task in pressure vessels is holding together, measured by three times the product of volume and pressure. Holding-together power may be exercised by sheets, working at points A (sphere) or E (cylinder) of Figure 3.10, or by filaments or tendons working at point B. Material working at B has only half the holding-together value of material working at A or two-thirds that of material working at E (58 per cent of that working at E on the von Mises criterion of shear strain energy). Nevertheless, in very large high-pressure (large-task) vessels, such as those housing gas-cooled reactors, filament-strengthened vessels are cheaper than steel shells. So much cheaper are they that a different form of design with all the coolant circuit within one pressure vessel becomes viable, eliminating the difficult thermal-expansion problems mentioned in Section 7.4. (This solution also involves an example of separation of function, referred to in Section 2.12.) A contributory factor in the superiority of the prestressed reinforced-concrete vessel is the high value:cost ratio (for material at point B) possible in steel-wire reinforcing tendons. But above all, the sheer size of the task favours the tendon-strengthened vessel in the following way.

The amount of material required for either type of vessel is roughly proportional to the task, or to pL^3 where p is the gauge pressure and L a typical dimension, so that a crude algebraic expression for the cost should contain such a term. However, fabrication and erection costs exceed material costs and here an important factor is the thickness of the plates of which the steel shell must be made, which varies as pL. Thick plates become very expensive to weld, so that we can expect the cost per ton of the steel-work to rise fairly fast with pL. On the other hand, the placing, anchorage and prestressing costs of each tendon will not increase much with its length, so that the total for all the tendons will increase little faster than pL^2.

It is to be expected, therefore, that the cost per unit of holding-together power of the welded steel shell will rise with the task, while that of the reinforced-concrete vessel will fall. For the same reason, it is cheaper to subdivide a steel shell into a number of smaller vessels connected by ducts, the minimum size being dictated by the largest indivisible unit, the reactor itself, while the most economical arrangement with reinforced concrete is a single all-enclosing vessel.

Based on the designed ultimate pressure of 1155 p.s.i.g. (7.9 N mm^{-2}), the task of the Oldbury 'B' Vessel, of about 279 000 ft^3 (7900 m^3), is 1.67×10^{12} lbf in. (189 GJ). At the ultimate load of the anchorages the holding-together power of the tendons is about 4.1×10^{12} lbf in. (460 GJ). The discrepancy is chiefly due to the thickness of the walls, which adds to the effective task; the median surface of the walls encloses about twice the volume of the vessel itself. This ratio would be less in a more modern design.

8.5   Other cases

Although some other functions have costs roughly proportional to the task, the majority of cases are not so simple. For example, rolling-element bearing costs are not proportional to load, but are least for middling sizes, with higher unit costs in small and large bearings. Nevertheless, this non-linearity does not prevent the development of function costing; it just makes it rather more difficult.

Because of their relative simplicity, their production in large numbers, and the highly competitive and open markets in which they sell, components like bearings are easily costed at an early stage. Sometimes very similar functions occur both in such components and in complete machines. For example, hydraulic cylinders for use as actuators may be bought as off-the-shelf items but some presses incorporate similar cylinders. In such cases it is interesting to compare the cost of the standard component with the cost of supplying the same function in the complete machine, to the extent that it is possible to attribute part of the costs of the whole to that particular provision. If the function appears to be much more cheaply provided in the standard component, then there may be scope for reduction of costs in the complete machine.

It is most desirable that cost estimates at the conceptual stage should be available and should be reasonably reliable, for it is at that stage that savings are made or are irretrievably lost. For the most part, reliance has to be placed on experience and judgement, with perhaps some detailing and part-by-part costing in difficult areas, such as where unfamiliar constructions are involved. In some fields some companies have developed more systematic methods, but there is scope for a great deal of useful work on how to secure reliable costings, at an early stage and at low cost.

9 Various principles and approaches

9.1 Introduction

This chapter collects together various helpful ideas. They are of fairly general application but too vague to merit the title 'methods', and they are not dealt with at length.

9.2 Avoiding arbitrary decisions: degrees of freedom of choice

Every choice with which the designer is faced represents an opportunity, a *degree of freedom* in the design which can be exercised to improve it. An arbitrary decision represents a wasted opportunity.

There are two difficulties in avoiding arbitrary decisions; the first is to recognise when a decision is being made. That this is not always easy is shown by the following example.

In the early days of the aircraft gas turbine, it was impracticable to make the turbine disc and blades in one piece. (It still is, in most applications.) The practical way was to make each blade a separate piece and fix a set of blades to a disc; the preferred form of fixing was the fir-tree root. All this was sound enough. The arbitrary decision which passed unnoticed was that the fixing was put at a radius just slightly less than that of the inner end of the aerofoil, as in Figure 9.1(a). The result of this was that the disc rim was at a temperature little less than that of the blade itself, restricting the choice of disc materials to those having refractory properties nearly as good as those of the 'Nimonic' alloys used for the blades. This meant the use of austenitic steels which were expensive and dense, and were also weak and unreliable compared with the ferritic steels which might have been preferred were it not for the high rim temperature. A further disadvantage of the austenitic steel was a high coefficient of thermal expansion.

A great improvement was made by Rolls-Royce, who put the fixing at a smaller radius. The truncated wedge of blade material left between the aerofoil and the fixing, sometimes called an extended root, was hollowed out on either side to leave an I section (Figure 9.1(b)). By this means the maximum temperature in the disc was reduced to a point at which it was possible to use creep-resisting ferritic steels. The combined and roughly equal effects of lower temperature difference between hub and rim and lower coefficient of thermal expansion reduced the thermal

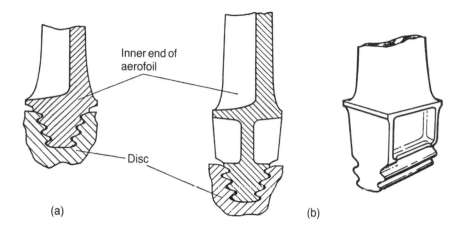

Figure 9.1   Extended root turbine blade.

stresses to about one-third in some cases (Section 4.2), and this together with the higher strength of the ferritic steel more than doubled the effective holding-together power (Section 3.12).

By this change, reductions in the weight of bladed discs of up to 55 per cent were effected. The fir-tree has to sustain a higher load on a smaller pitch length, but this was made possible by the better properties of the material of the female part. Savings on the cost of the disc forgings were offset by the greater length of stock required for the blades, and slightly more cooling air may have been needed, but the saving on total engine weight was a great improvement.

While it is easy to be wise after the event, the delay of almost a decade in arriving at this solution could scarcely have occurred had a sufficiently abstract view of the problem been taken in the first case, especially if this particular variable, the radius of the fixing, had been recognised as such. Notice the role of established ideas, the blade and the disc, and the example of steam-turbine practice, in obscuring the important freedom of choice.

Once the impracticability of making the wheel (bladed disc) in one piece is established, and the likelihood that it will be profitable to use different materials for different parts is recognised, then the abstract view of the problem is that of making one part (a wheel) in two materials, A and B. It is then automatic to regard not only the natures of A and B but also the position of the boundary between them as a matter for choice. It is for this sort of reason that it is advocated that the designer should continually increase the level of abstraction at which he looks at his problems, at least to the point where he is tolerably sure that the process has become unprofitable.

Exploiting a freedom of choice

The second difficulty in avoiding arbitrary decisions is more conspicuous in the embodiment than the conceptual stage; it is that of deciding how to take advantage of the degree of freedom of choice once it has been recognised. Sometimes it is difficult to see any way in which it can be useful, and sometimes it is difficult to be sure that its greatest potential use has not been overlooked.

As a rather trivial example, consider the location of six bolt holes on a given pitch circle in a flanged joint. They may be made equally spaced, and this is the best arrangement from a structural point of view. This still leaves a degree of freedom in the angular position of the set of holes, a detail which may well be determined by studying the accessibility of the bolts for insertion or tightening. On the other hand, it may be desirable to make the spacing uneven to prevent assembly in an incorrect angular position.

At an earlier stage, it was necessary to decide on a number and size of bolts and a pitch circle diameter. This problem is generally a structural one, though stiffness to ensure sealing may well be the critical requirement rather than strength. If the pitch of the bolts is high, internal pressure may separate the surfaces between bolts enough to unload the gasket to a point where it is blown out, or fails to seal. In such a case the number of bolts can be reduced if the structure between them can be stiffened. Feilden, by stiffening the cylinder-head structure of a diesel engine, was able to hold 2000 p.s.i.g. ($13.8\,\mathrm{N\,mm^{-2}}$) with four $1\frac{1}{4}$in. (32 mm) bolts where before seven $1\frac{3}{4}$in. (44 mm) bolts had only been able to hold 800 p.s.i.g. ($5.5\,\mathrm{N\,mm^{-2}}$) [37].

9.3   Mathematical models

It might be thought that the mathematical model is too familiar a notion to require special mention in a book of this sort. Nevertheless, designers are apt to restrict their use of mathematical models to those they have met in textbooks or technical papers. In the cylinder-head design problem just discussed, the failure to use a simple mathematical model was a root cause of the original weakness. The good designer will try to devise mathematical models of his own, even in the least promising situations. As an extreme example, it is perhaps fair to mention Lanchester's pioneer work on the operational theory of warfare.

Beds

To illustrate the application of mathematical models to an unlikely field, the design of beds will be considered—beds for human beings, not machines. Market research may return rather subjective statements of preference—'comfortable but free from sagging' or 'soft yet firm'. How can we find useful measures or concepts (Section 4.5) that may be related to these qualities?

For the bed we substitute a mathematical model in the form of an elastic foundation. To simplify the treatment, it will be restricted to a thin vertical slice of bed taken parallel to its length, so the problem is in two dimensions only, length and depth. An axis of x is taken in the unloaded level surface of the bed, and an axis of y downwards (Figure 9.2). In the unloaded condition the surface of the bed is given

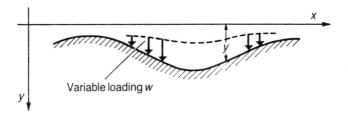

Figure 9.2   Loads on idealised bed.

by $y = 0$, and under a loading w per unit length, where w is an arbitrary function of x, it is given by

$$y = F(w, x)$$

It is more convenient for our purpose to invert this relationship to give w in terms of y,

$$w = f(x, y) \tag{9.1}$$

Now this function, f, describes the stiffness of the bed surface, and it is the designer's task to ensure that it has as nearly as possible the most suitable form, whatever that may be. Let us see what can be done with the excessively general and unintelligible equation 9.1.

Firstly, let us assume that the bed is uniform throughout its length and that the ends do not greatly differ in their properties from the rest. Then a shift of the origin of x can make no difference in the relationship of w and y, so that x cannot occur explicitly in equation 9.1, which may be rewritten

$$w = f\left(y, \frac{dy}{dx}, \frac{d^2y}{dx^2} \cdots\right) \tag{9.2}$$

Secondly, let us assume that the bed is linear; then equation 9.1 can be rewritten

$$w = k_0 y - k_2 \frac{d^2y}{dx^2} + k_4 \frac{d^4y}{dx^4} - \cdots \tag{9.3}$$

Odd-order derivatives cannot occur because their sign would reverse if the direction of x was reversed. If

$$y = y_0 \cos\left(\frac{2\pi}{\lambda}x + \phi\right)$$

substitution in equation 9.3 yields

$$w = \left[k_0 + \left(\frac{2\pi}{\lambda}\right)^2 k_2 + \left(\frac{2\pi}{\lambda}\right)^4 k_4 + \ldots\right]\cos\left(\frac{2\pi}{\lambda}x + \phi\right)$$

$$= y_0 c(\lambda)\cos\left(\frac{2\pi}{\lambda}x + \phi\right) \tag{9.4}$$

where $c(\lambda)$ is the expression in square brackets. The right-hand side of equation 9.4 can also be written in the form

$$w_0 \cos\left(\frac{2\pi}{\lambda}x + \phi\right)$$

where

$$w_0 = y_0 c(\lambda) \tag{9.5}$$

Then equation 9.4 can be interpreted as follows: a sinusoidal load of amplitude w_0 and wavelength λ produces a sinusoidal deflection of the same phase and wavelength and of amplitude

$$y_0 = \frac{w_0}{c(\lambda)}$$

In other words, $c(\lambda)$ is the stiffness to a loading of wavelength λ. We can describe

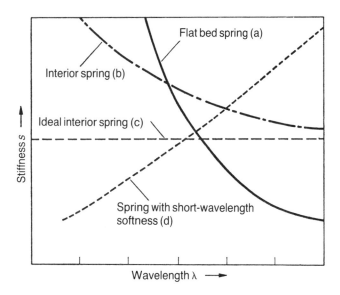

Figure 9.3   Stiffness wavelength relations for beds (hypothetical).

the characteristics of a particular bed structure by the function c. If s increases with λ, the bed is stiffer to long-wavelength loadings than to short-wavelength ones.

Now let us think what this means in terms of comfort. To support the sleeper with his spine free from flexure, a relatively high stiffness is required at wavelengths of about 1 to 2 m. At wavelengths of 0.5 m or less a lower stiffness is desirable so that local bumps on the sleeper, like arms, shoulders and haunches, can sink easily into the bed without developing high local pressures; many people will have felt at some time they would like a hole in the bed to slip their arm into, if only that hole could always be in the right place.

Flat bed springs, hammocks and such approximate to the case where only k_2 in equation 9.3 has a significant value, so that c is proportional to λ^{-2} (Figure 9.3).

The separate coil springs of the conventional interior spring mattress ideally have all the ks in equation 9.3 zero except k_0, so that c is constant. The covers of such mattresses and 'spring bases' are generally sufficiently inextensible to introduce a measure of k_2 however, so the practical curve of c against λ would probably show a considerable downward trend, while what is needed is an upward one—'short-wavelength softness' might be our slogan. It seems to the writer that some designs of continuous spring may possess in themselves a measure of short-wavelength softness, but this is probably swamped by cover tensions. Other forms which are potentially good are some older types of foam mattress and some recent types of slatted construction. Water beds approximate to the uniform c of graph (c).

One final point is that the mathematical model enables us to change our viewpoint (Section 4.10). Having decided on the curve of c against λ we want (perhaps as in Figure 9.3) we can derive the deflection under a point or very local load, as in Figure 9.4. Notice that the 'shorter-wavelength softness' bed rises up on either side of the loaded point, a feature that at once suggests ways in which such a characteristic might be achieved.

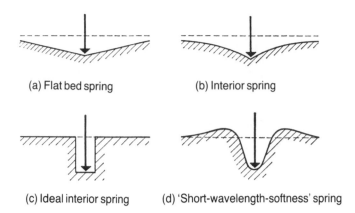

(a) Flat bed spring   (b) Interior spring

(c) Ideal interior spring   (d) 'Short-wavelength-softness' spring

Figure 9.4   Deflection under very local load of beds of Figure 9.3.

9.4   The search for alternatives

The importance of looking for alternative ways of doing things is well recognised and has received attention elsewhere in this book. Even in a combinative treatment, however, possibilities may be overlooked. A great help would be a comprehensive list of known engineering means, a standard work on the engineering 'repertoire' (Section 2.3).

Books of various kinds exist, indeed, but they are generally limited in scope, more detailed than is necessary for the present purpose and not abstract enough in their classification; they tend to overlook the kind of alternative the designer overlooks, through their concern with types of device, rather than purposes.

As a simple example, suppose we look through a list of materials to find one of high thermal conductivity; it is unlikely we shall find quoted the extremely high effective values given by various forms of natural convection—a single phase of sodium or two phases of water in a high centrifugal field, the steam tube in a baker's oven, a heat pipe where capillary action returns the liquid phase or perhaps a magnetic fluid in a suitable magnetic field. Yet one of these might suit your purpose.

One particular alternative that can always be considered and occasionally leads to a profitable idea is to dispense with a function altogether. As an example, a small team working under the writer's guidance on the design of a very small heat engine using the Rankine cycle found that one of the difficulties was the feed pump. They finally put the condenser above the boiler and arranged a float valve to lift and put the condenser in direct communication with the boiler when most of the working fluid had accumulated as liquid below the condenser. This liquid ran down into the boiler, now depressurised, the valve was closed, and the turbine started again when the boiler boiled once more. The extra irreversibilities and intermittent operation were no great disadvantage in this particular case, but the elimination of the feed pump was a substantial gain.

Another important type of alternative easily found by a systematic approach is that involving *inversion*, either in the narrow sense as applied to academic mechanisms or in a wider sense. The oscillating engine used a mechanism which was an inversion of the slider-crank chain, obtained by making the connecting rod (coupler) the fixed member. The star gear is an inversion of the simple epicyclic. But more important are inversions in the wider sense, such as reversing the male and female roles between two parts which screw together, or making a shaft run with its bearings inside it on a central pin, instead of with its bearings outside it in fixed housings. Usually the designer will hit on the best form first time, and the inversions can be discarded after scant consideration, which they should receive, however. A simple but helpful example is given in reference 38.

9.5   Logical chains

A useful device is the logical chain of reasoning, leading often to a unique practical solution. Sometimes this solution will appear eminently satisfactory, and sometimes it will seem unduly clumsy or expensive. In either event it is worth testing the chain link by link in the hope of finding some fault in the reasoning; the good solution may be bettered, and the poor solution proved not to be unique.

As an example of an idea originating in a logical chain, the writer was asked some years ago what use could be made of the cold sink available when liquid natural gas shipped in from Algeria was evaporated and warmed to ambient temperature preparatory to use. Various suggestions had been made that it be used as the cold sink of a heat engine. The first step in gaining insight is to recognise that the potential money value of one joule of available energy of a liquid gas as energy is less than that of one joule of available energy in a common fuel, because of the special, and hence expensive, plant needed to convert it. The money cost of producing the available energy (or cold, for short) is relatively high because of the special, and hence expensive, plant required to liquefy the gas. It follows that if a really profitable use is to be found it must be as *cold*, not simply as *available energy*.

The next step is to enquire what existing uses there are of this quality of cold in such large quantities, just as might be asked of any material commodity. The simple answer is easily missed; the same quality and quantity is used in liquefying the gas for shipment. If the cold could be shipped back, it could be used to liquefy more natural gas.

The only way to ship the cold back is in some substance; the most attractive possibility is a liquid gas, but the choice is limited since it must either be a genuine return cargo or something of very low cost. The only gases which are cheap enough are those present in large proportions in the air, and safety considerations rule out oxygen. We are left with nitrogen as the only possibility.

Let us see where we have arrived. A ship loads LNG at A say, and sails to B where it unloads LNG and takes on liquid nitrogen. During the interval before the arrival of the next ship a conversion plant at B vaporises and warms LNG and separates and liquefies atmospheric nitrogen—the separated oxygen gas can perhaps be sold.

By the time the next ship arrives, there is a fresh cargo of liquid nitrogen ready. At A there is a plant which carries out the reverse process, liquefying natural gas using the cold in the liquid nitrogen, which emerges at atmospheric temperature and pressure.

Because of evaporation en route and irreversibilities in the plants, the process cannot be self-sustaining, so that at one or both ends there must be a substantial power input. The amount of power needed depends on the difficulty of matching nitrogen to methane, a problem rather like that of the dual-pressure steam cycle in Section 5.7. It transpires that the difficulty is great, and this together with the loss of valuable flexibility in the rate of vaporization of LNG at B make the idea less attractive. Nevertheless, it was worth studying.

The most important step here was the recognition of the potential use of the cold arising from giving the right answer to the question: 'Who uses this quantity and quality of cold?' A question and answer approach is often helpful, and valuable questions often seem childish at first. Another important step was recognition that the cold could only be of much value as cold.

As another example, consider the problem of the slow heating up of the rings of electric cookers. The rings are resistance heating elements that have to be robust since they are in contact with the pans and subject to fairly rough usage. They have also to be electrically safe, though exposed, and to provide a sufficient surface area for heat transfer to the pans. These requirements impose a tight lower bound on the volume and hence the water equivalent of the rings, and this in turn means a long heating-up time.

This logical chain has at least three weak links. Firstly, it is not essential to use resistance heating outside the pan – we could use microwaves, for example. Secondly, it is not necessary to have the rings in contact with the pans—we could use an intermediate heat transfer agency. Thirdly, the long heating-up time is not absolutely necessary given a high water equivalent, since we might be able to boost the power to begin with. None of the three seems very promising but a combination of the second and third is attractive.

Wave energy converters

An interesting case is that of wave-energy converters. It was seen in Chapter 2 that the function of providing the necessary reaction against the wave force on the working surface could be supplied in one of three ways: fixing the device to the seabed, opposing one wave force with another, or using inertia. Now, as remarked in Chapter 2, the use of the inertia of a very large mass would be prohibitively expensive, and so this option was rejected. However, this link of the logical chain may be breakable.

One of Sherlock Holmes' favourite axioms was, 'When you have excluded the impossible, whatever remains, however improbable, must contain the truth!' Substituting 'solution, if there is any' for 'truth', this axiom becomes a valuable aid to the designer. Let us examine its applications to wave energy.

The costs of fixing to the seabed or balancing one wave force against another being high, perhaps we should look again at inertia. The reaction developed by accelerating a mass is the product of the mass and the acceleration, and by using a large acceleration we can develop the required force with an acceptably small mass. To see how this may be done, look at Figure 9.5, in which a buoy similar to that in Figure 2.11 is shown: a wave crest is approaching from the left, so the buoy is rising. It has no attachment to the seabed whereby this rise can be partially resisted, so extracting the energy, but it contains a mass attached to a compressed spring and restrained by a catch. The spring is pushing down on the bottom of the buoy, and up on the mass and hence the catch, so the upward and downward forces balance. If the catch is released, however, the upward force is temporarily removed and only the downward force is left and will resist the rise of the buoy as required. When the

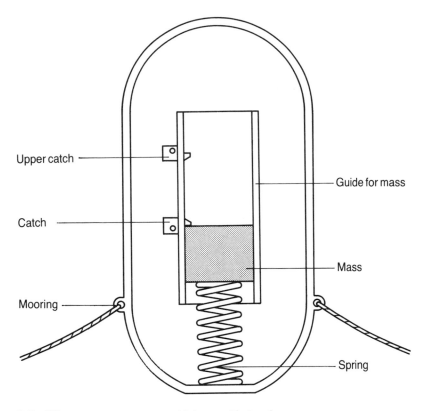

Figure 9.5   Wave energy converter with internal balancing.

mass momentarily stops, it can be secured by the upper catch shown and is then available to provide a reaction to extract energy on the downward stroke. A subtle control system is needed to release the catches at the right moment so as to extract the most energy from irregular waves, but that is well within the state of the art. The real problem is the spring, which is easy enough in a small model, but daunting at full scale. Energy is extracted from the moving mass, e.g. by making it pump air through a turbine. Fixed catches will not do, otherwise the device will not start working: an arrangement must be provided which can hold the mass wherever it stops, but that is not too difficult. If we can overcome the spring problem, this will be the most promising WEC yet.

Incidentally, in Section 4.6 it was remarked that 'reactionless drives' and 'anti-gravity' machines were a modern equivalent of the perpetual motion machine, but this WEC does not fall into that category because the time average of the reaction that it produces from within itself is zero.

9.6   Past practice and changed circumstances

Every designer automatically and necessarily draws very largely on what has been done before. It is difficult to do this without accepting as essential things which are not, making provision for circumstances which do not arise, or adopting practices which are no longer, or never were, the best available. Some defence against these sorts of error is provided by a systematic review of the logic behind the scheme in question and a diligent search for viable alternatives, as briefly discussed in the last few sections.

A designer tackling a problem outside his own field often has a freshness of approach and a freedom from preconceived ideas that is of great value. It has been argued that this aspect is so important he should deliberately avoid acquainting himself with past practice, but start absolutely from first principles as if he were the first man to work in the field in question. The writer feels this is too extreme an attitude, and that it is possible to retain the advantages of starting from scratch while securing the valuable insight which can often be gained from a study of past practice. Since gifted designers are often of great individuality, not to say perversity, it is not likely that they will be blinded to alternatives by orthodox practices without a long period of habituation.

9.7   Brainstorming

A useful technique is that called 'brainstorming' [39], in which a group of designers sit down to produce as long a list of 'solutions' to the problem in hand as they can in a given time. The essence of the method is that no analysis of the 'solutions' is attempted at the time, though the writer's limited but encouraging experience suggests that this rule should not be strictly adhered to.

Each designer (or other brainstormer) contributes a tentative idea as soon as it comes into his head, and this will often stimulate a chain of further suggestions from his colleagues, generating a kind of momentum which can easily be lost if too much time is given to any particular solution. The end product of the 'session' is a host of tentative ideas, some of which may be valuable.

It is usually said that no criticism should be allowed, either because it is inhibiting or because it breaks the flow of ideas. The writer, however, agrees with Dixon [40] that short objections to suggestions often prompt the next one, and in any case, the members of the group need to be got into a frame of mind where they are not inhibited by criticism.

A tape-recorder is useful—note-taking slows the process up for at least one member. However, it is useful for the chairman to make notes of any point at which there was a tendency for the development of ideas to branch. Then, when the chosen branch peters out, he can bring the session back to the alternative line. In a thriving brainstorm there are always too many directions to follow all at once, and a 'reprise' will often restore the momentum. Above all, no-one should be allowed more than, say, three sentences at a time (unless his idea is particularly complicated).

Reference 40 is excellent on this topic.

9.8   Use of solid models

There are some design problems in which solid models can be of the greatest help, and even the advent of powerful computer graphics systems is unlikely to change that situation. The most recent major case in the writer's experience (1984) was that of a vehicle door (Leyland Vehicles Ltd.) in a study made through a Science and Engineering Research Council Teaching Company. The door consists of a few major components and many smaller ones fitted together, the shapes are subtle and the interactions are complicated and difficult to grasp. In such a case it is invaluable to be able to hold models in the hands and fit them together. The first models were made of card, but later it was found worthwhile to vacuum form $\frac{1}{4}$ scale models in thin plastic. One benefit of this was that many part models could be made at little cost and cut about, stuck together and modified in various ways. It is difficult to see how this particular design could have proceeded as well without the models.

Usually the models are less well made, as for example, simple linkages cut from card and using drawing pins as pivots. Linkages can readily be designed on the computer, but it is often helpful to have a cardboard model in addition. Casings can be designed and stressed on the computer, but if the stresses are unacceptable in spite of scantlings that look substantial, a cardboard model flexed in the hands may provide insight into why it is so poor structurally, and how it may be improved.

Sometimes a solid model may prompt ideas, and suggest alternative forms better in function, more convenient in use, or easier to make. Drawings may do this too, but the model may be a more powerful stimulant because it is more easily grasped.

Finally, as the industrial designer knows well, solid models often provide the most immediate and vivid way of communicating ideas, whether to other members of the design team, to superiors, or to clients.

It is difficult to anticipate when solid models may be of help. Besides the cases mentioned, the writer has used them for a root fixing for a ceramic turbine blade, where communication was the chief advantage, and for two novel liquid natural gas tanks where the object was to help form a judgement on their practicality by visualising them better.

9.9   Some maxims for designers

At this point it is convenient to summarise some of the advice that has already been given, in some succinct maxims that should be useful. Their very brevity may make them more valuable, because every designer has his own way of thinking and will want to adapt any such general advice to it.

1. Increase the level of abstraction at which the problem is formulated.
This is perhaps the most useful single idea in design, and much of this book exemplifies it, from Section 1.9 onwards. Removing the problem to a higher level of abstraction often suggests different solutions, and sometimes they are better ones. Above all, many preconceptions and arbitrary decisions are left behind and 'of course' is replaced by 'why?' or 'why not?'.

2. Make simple functional analyses and tables of options (as in Chapter 2) or better, make more elaborate ones and prune them down as the work develops and it becomes clearer which are the key functions.
There are few problems where this approach is of no help at all, but it is most useful where the thing to be designed is new of its kind, or (and this should not be overlooked) in the design of relatively small and simple components. It is sometimes forgotten that detail design consists largely of the conceptual design of

components, and much of what applies to the design of the whole can be used in the design of parts.

3. Press to limits.
Suppose a design has been reduced in cost by making it longer and thinner, then we should ask whether it would be even cheaper if it were still longer and still thinner. If a car engine has been improved by making the cylinder 'over-square' (i.e. the stroke less than the bore), would it not be better still if were more over-square? Generally, the answer to such questions is that, beyond a certain limit, some disadvantage associated with the change begins to increase rapidly, while the original advantage yields diminishing gains. But, whenever a beneficial direction of change has been found, it should be pursued until the limit has been reached.

4. Aim for clarity of function.
This may seem like unnecessary advice, but much design is confused and elaborate where it might have been clear and simple. Sometimes this is because of a reluctance to start again when difficulties arise, and a preference for adding devices to overcome failings, which leads to what has been called the 'Christmas tree' style (Section 2.11). As an example, in a gas turbine with a free power turbine, a failure of the output shaft will leave all the output torque accelerating the free turbine rotor, which may overspeed and burst. This risk can be avoided by sensing the rise in rotor speed and cutting off the fuel. However, in one design of such a turbine, where the output gear box was fitted with a torquemeter, it was thought preferable to use the drop in output torque as the indicator of shaft failure. The first complication was that when the turbine was not running the torque was zero, shaft failure was indicated and the fuel was cut off: it was therefore necessary to suspend the safety device when the throttle setting was zero, so that the engine could be started at all. Also, since not all the torque would disappear if the output shaft failed, the turbine had to be cut off if the torque did not disappear, but merely dropped below a certain threshold. Moreover, this threshold had to be variable according to turbine speed, which when running steadily depended on throttle setting: accordingly it was arranged that the threshold was altered by altering the throttle setting (and also by changes in altitude, since this was a helicopter engine). Unfortunately, if the turbine was idling and the throttle was suddenly opened up, the threshold value was raised by moving the throttle to a torque higher than that produced by the idling engine, which immediately cut out. To overcome this difficulty, a delay was arranged between moving the throttle and changing the threshold value. Further complications arose, until at last it was proposed to fit a speed-sensing element to the power turbine which would control the threshold setting. But the whole complicated box of tricks could have been replaced by a speed sensor on the power turbine which would shut the engine down if it overspeeded.

One word of caution is perhaps needed with regard to the principle of clarity of function. Taken too far, it would lead to the shunning of the kind of solution described as 'hybrid' in Chapter 2, yet there are many situations in which hybrid designs give the most economical solutions. This is particularly true where the functions themselves are not exacting, and considerations of manufacture and assembly dominate, as in consumer goods and vehicles.

5. Apply kinematic and elastic principles of design.
These ideas have been dealt with in Chapter 7, but it is worth stressing their great

importance and the troubles that are likely to attend their neglect. In instruments in particular a rather 'pure' approach pays. For example, in load sensors purely elastic designs are to be preferred, but in less critical cases there is often a temptation to use kinematic pairs to reduce costs. Resist it, for the troubles which result tend to be greater than expected and are only likely to be justified if the saving is large.

6. Exploit materials and manufacturing methods to the full.
Every designer will naturally intend to do this, but nevertheless it is worth considering deliberately and systematically whether it is being done, at several points in the design process, but particularly at the start of the embodiment stage. It is important to keep abreast of new developments, and here regular reading of the technical journals is the greatest help. For instance, at the time of writing (1983) developments in adhesives over a long period are coming to their fruition in very economical and convenient joining of large structural pressings. Polymers generally are an area where there are many clever uses of manufacturing tech- niques. A well known example is boxes in which box, lid and hinge are moulded in one, the material properties of the thin hinge section becoming more suitable with flexure. It is easy to overlook possibilities like building-up a local area of a mild steel part where high hardness or wear resistance is required by welding using a high alloy rod, or simplifying the moulds for a die-cast part by making it in a form which is locally bent after casting to achieve the required final shape.

7. Make and break logical chains (Section 9.5).
Both these steps, the making and the breaking, are of value, and sometimes the inability to carry them out is even more important. Most good designs can be justified by a fairly tight line of reasoning, after they are arrived at, if not before. If no such line can be established, it is difficult to have confidence in a design, and more conceptual work is indicated. On the other hand, to do better than a design which can be so defended, we must look for the weakness in the argument. In the case of the wave energy converter, this was that the inertia need only be large if it was a *passive* inertia.

8. Ask naïve (stupid?) questions.
However obvious the answer may seem, the designer should ask questions of himself and others: 'Why is it necessary?'; 'Do we have to be able to take this part off again?'; 'What would happen if we didn't bother?'; 'Would it matter if it broke?'. An example was given in Section 9.4, where the answer to the naïve question, 'Do we need a feed pump in this particular Rankine cycle?' was 'No'. This maxim is closely related to maxim 7, and is part of a sort of systematic incredulity and iconoclasm that the designer should cultivate.

9. Count your blessings as well as your woes.
Because functional design is always difficult, it is easy to overlook the aspects of a problem which are less difficult than usual. For example, if a device has to function only in an emergency, or for a short time, various considerations which would normally apply are less important: efficiency may be relatively insignificant, cooling may not be required, higher stresses may be tolerated, and so on. In one case, an energy-saving device on a mechanical power transmission was being studied. One simple concept had the defect that it would not work with cold oil, but this was of no importance because the transmission still worked (but without the saving) and the oil soon warmed up: the savings with the simple design were very nearly as great as with a much more elaborate scheme.

These nine maxims I would recommend to any designer as worth his serious consideration, but they may be of particular help to the inexperienced. Old hands at design probably follow some of them already in their own versions no doubt, but may still find value in seeing them set out, or be prompted to further advances of their own.

Questions

Q.9.1(2). Figure 9.6 shows the geared elements of an automatic transmission. Gears 7 and 8 are welded together and are free to turn on the pin through their centre. Items 1 to 5 are initially free to rotate about the centre line of the whole system except in so far as they are constrained by the various meshes.

Satisfy yourself, by enumerating spatial degrees of freedom, that the system has two, so that one further constraint is required to transmit a drive.

The input is always at 1, and in forward gears the output is always at 5. These gears are obtained by applying the following additional constraints.

    First, hold 2   Third, hold 3
    Second, hold 4   Top, lock 4 and 5 together

Show that there is only one degree of freedom of choice in fixing the set of gear ratios, i.e. that if any gear ratio (except top, which is 1:1) is fixed, all the others are fixed.

Find the concealed arbitrary decision in this design; i.e. find the change in geometry that can be made without altering the topology or essential nature of the gearing system that will increase the number of degrees of freedom to two, so that two gears (other than top) can be freely chosen (within the limits imposed by practical numbers of teeth). (This was the basic gear system of the Automotive Products automatic transmission used by BMC. The concealed degree of freedom was observed by the designers, who decided not to use it for reasons given in

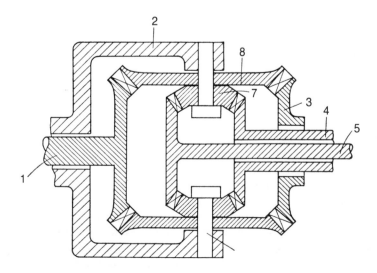

Figure 9.6   Gearing for automatic transmission.

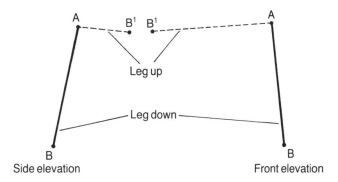

Figure 9.7   Undercarriage leg geometry.

reference 41: although superficially these reasons are excellent, the writer believes they conceal another arbitrary decision, that of Question 9.5, in effect.)

Q.9.2(3). Figure 9.7 shows two views of an aircraft undercarriage used to illustrate a problem in descriptive geometry, which was as follows: AB represents a strut of the undercarriage in the down position: it is hinged at A. In the retracted position AB is at AB'. Show the axis of the hinge pin in both views.

Find the arbitrary decision inherent in this question. To what sort of use might the suppressed degree of freedom of choice be put?

Q.9.3(2). You are a member of the committee for the decimalisation of the currency in Ruritania. None of the old coins will correspond to convenient numbers of the new hopes (100 hopes = 1 zenda). Putting aside all preconceived notions and given that the smallest coin is to be 1 hope and the smallest note 1 zenda, what denominations of coinage would you suggest?

Q.9.4(3). Devise a simple mathematical model for the springing of the rear wheel of a car which will fit solid axles, independent suspension, and independent suspension with an anti-roll bar.

Q.9.5(3). It is sometimes stated in works on gearing that the number of teeth in the planet gears of a simple epicyclic gear must be half the difference of the number of teeth in the annulus gear and the sun pinion. This is a fallacy. Why?

Answers

A.9.1. The axis of the gear cluster 7 + 8 need not be perpendicular to the axis of the system.

A.9.2. It is arbitrarily decided that the hinge axis must be perpendicular to the strut. In fact, any hinge line through A and lying in the plane bisecting BB' at right angles will do, i.e. there is a degree of freedom of choice which might be used, for example, to make the wheel tuck flat into the wing.

A.9.3. It is an arbitrary decision that the denominations must be factors of 100. The most convenient choice of three denominations between 1 and 100 is 3, 10 and 30, requiring on average only about 80 per cent as many coins giving change and smaller numbers of one denomination to count, compared with 2, 5, and 20.

A.9.4. Turn the car upside down and apply a downward force F to one wheel. Suppose that that wheel descends a and the other one b. Then a/F is the *direct flexibility* and b/F the *cross-flexibility* and these two together define the properties of the suspension (in a simple treatment). The cross-flexibility is negative with a solid axle sprung inboard of the wheels, zero with an independent suspension and positive with an anti-roll bar on an independent suspension.

A.9.5. The fallacy arises from the arbitrary assumption that the pitch circles of the planet gear in its meshing with sun pinion and annulus must be the same circle. There is no reason at all why this should be the case—indeed, it is a special virtue of the involute gear that the pitch circle is not fixed. The gears in Figure 7.7 have numbers of teeth 24, 47, and 120.

10 Conclusion

10.1 Resumé

The preceding chapters have dealt with various methods and approaches that may be helpful in conceptual design problems. Some of these, like the combinative studies of Chapter 2, are of general application. Others, like the identification and treatment of generic types of problems such as matching (Chapter 5) and disposition (Chapter 6), are only of use in special cases; nevertheless, as the list of topics in Chapter 5 will show, the special cases are so frequent that the approach is of great value. The insight-developing techniques of Chapter 4 are strongly recommended, and the writer believes that the investment of time in such exercises will be amply repaid in better designs more expeditiously arrived at. The advantages of rough calculation of orders of magnitude and 'quick sums' in particular cannot be over-emphasised; the optimisations dealt with in Chapter 3 are of this sketchy sort. The space available would not allow of fuller treatments, but in any case these simpler exercises yield perhaps nine-tenths of the possible benefits at a few per cent of the cost. Of course, the heavy calculations that fetch in the remaining tenth are still very worthwhile whenever the sums of money to be spent are large.

Chapter 7 (Kinematic and elastic design) differs from the rest in the restricted physical nature of its subject matter, which nevertheless is of such importance as to warrant its inclusion. (The writer abandoned a similar chapter on structures when it exceeded a third of the length of the present book.) Most engineering artefacts of any size or importance warrant a consideration of the conflicting merits of kinematic and elastic alternatives somewhere in their design.

Chapter 8 (Costs) was even shorter in the First Edition and remains rather inadequate, but the writer still expects this to become an important field of knowledge and a major support of design management in the near future. The ideas of Chapter 9 will be familiar to many readers, but it is hoped they will find the examples interesting and valuable, and the whole a useful reminder.

10.2 Checking and evaluation

Every designer is only too familiar with the pitfalls that beset him—an exceptionally treacherous one is described in Section 8.3. The best protection against them is a complete review of all the processes of manufacture, assembly, operation, and maintenance of the product, and most of us can benefit from a prompting check-list; here is one.

Function

Does it work?
Does it meet the specification?
Will threads seize owing to heat, rotors seize owing to transient thermal expansions, bearings overload for the same reason, bearings fail before oil reaches them or before they generate a film, will bearings brinell, will the starting torque be excessive, will thermal distortions or loads misalign bearings, will coolants freeze, will air, water, oil, dust, explosive mixtures, swarf, etc. collect in unsuitable places, will seals or sumps produce excessive drag, parts fret, nuts slacken, threads strip after repeated tightening, lubrication or fuel systems fail to function in certain positions, unequal tightening of bolts cause distortion, balance be lost due to stripping for assembly, bits fall into inaccessible places? (they will!).
Can it be got into its required location?

Manufacture

Can it be made?
Are the materials available, can they be treated and machined as required?
Can it be assembled? in only one way? without the help of periscopic-eyed dwarfs with socket-headed fingers?
Can the parts and the assembly be handled and transported? Are lifting eyes, lashing points, etc. required?
Are special tools needed for assembly or maintenance?
Are there alignment problems, requiring special equipment?
Are castings, forgings, welds, etc. within the capacity of the shop?
Is a practical sequence of heat treatment and machining possible?
Is any machining after assembly needed? and, if so, is it possible?

Maintenance

Have inspection covers been provided?
Can bearing clearances, etc. be checked?
Have suitable instrumentation, oiling and greasing provisions been made?
Can parts likely to need replacement be replaced *in situ*?
Can worn parts be salvaged, if necessary with the replacement of other parts with oversize ones?
Can flushing and cleaning be done as required?

Strength

Have all loading conditions been considered?
Have all modes of failure been considered, i.e. overload, fatigue, creep, brittle fracture, fretting corrosion?
Have vibration, thermal stresses, impact loads been considered?

Have effects reducing strength been considered, i.e. temperature, tolerances, finish, chemical nature of environment, corrosion, variation in material properties, internal stresses, welding, stress raisers, wear, effects of fitting or maintenance?
Are differential expansions allowed for? transient as well as steady?
Could mistakes in operation in service have dangerous results?

Safety

Is it safe?
Have any safety regulations been complied with?
Are warning notices required? interlocks? warning devices? fire-extinguishing devices? overload devices? overspeed devices?
Is it electrically safe?
Has drainage provision been made?

The process of evaluation has been covered fairly well in discussing design itself, since many of the methods involve evaluation in their execution. Nevertheless, no mention has been made of the financial calculations needed to carry out many of these evaluations properly, and indeed this is too large a topic for this small book. References 42 and 43 are suitable reading for the designer who is concerned with products where the proper techniques make a substantial difference—that is to say, almost anything where the total value of the product to be made is sizeable and the life of the plant or production line is longer than about a year.

Finally, an important aspect of evaluation is the cost of the designing—it is not worth expending great ingenuity to reduce the cost of a one-off product by £100 or so. On the other hand, the saving of two pence on a car door lock of which perhaps twenty million will be made will justify the employment of four good designers, each earning £10,000 per annum, for five months on studies in parallel and series (the rough calculation allows for overheads, associated detailing and costing work, risks and a usurious rate of return). The writer suspects that the proportion of the cost of most products that goes into design is much too small, but of course a limit is set by the amount of developed talent available. This is why the industrial designer is invading the field of engineering design proper. He has brains, creativity, and a sketchy knowledge of engineering, and he is rushing in to fill the vacuum left by the lack of suitable engineering graduates.

Questions

The following questions are intended to provide practice in the application of the ideas discussed in the rest of this book, e.g. the first question calls for ideas from Chapters 2 and 8. Most of them are rather difficult and will require long and deep thought, and perhaps some help from the brief answers which are supplied. A few of the questions are very difficult, but this is in the nature of conceptual design, which is among the most difficult of all intellectual exercises.

Q.10.1 A vertical take-off and landing (v.t.o.l.) aircraft has a wing which can be rotated so that its leading edge points upward. Four propellers ranged along this wing are driven by four gas turbines and interconnected by shafts so that all rotate at the same speed and continue to do so even if one turbine fails. Study possible arrangements of the engines, gearing and drive shafts, if the propeller speed is to be 1/20th of the turbine speed.

Q.10.2 In the optimisation of ship speeds, should the value of the cargo enter into the calculations?

Q.10.3(3). In a plant for desalinating seawater by forcing part of it through a membrane which does not allow the passage of salt, the seawater is raised to a pressure of $15 \, N \, mm^{-2}$. The flow through the membrane is 0.40 of the flow into the plant, the remainder emerging as brine. To economise in energy consumption, the brine is passed out through a turbine, which helps to drive the seawater pump. If both pump and turbine have efficiencies of 0.80, what is the energy consumption per kilogram of fresh water? Can you suggest any means of reducing this figure, without increasing the pump and turbine efficiencies and keeping the yield of fresh water fixed at 0.40 of the seawater intake? Reflection about whether the pV term in the expression for enthalpy is 'real' energy may help.

Q.10.4. The hopper shown in Figure 10.1 is used for a material that may be regarded as a liquid of density $600 \, kg \, m^{-3}$ as far as the forces it exerts on the structure are concerned.

The doors which close the bottom are hinged along AB and CD; the material is to be mild steel, with an allowable maximum shear stress of $60 \, N \, mm^{-2}$.

(a) Estimate to ±20 per cent the weight of steel required using two- and three-tiered structures. Do not include the legs below ABCD.

(b) The doors are to close automatically after releasing the contents of the hopper. Suggest means of achieving this. Should the means be located at both ends? Try to make your arguments quantitative.

Q.10.5. In machine tools there are commonly several relative degrees of freedom between the work and the tool, some concerned with the generation of surfaces, one with the feed, and some with positioning. For example, in a lathe, when used for turning a cylinder, spindle rotation and sliding generate the cylindrical surface while the traverse supplies the feed. In a radial drill, rotation of the spindle and the feed generate the surface, while swinging the arm and moving the head along the arm give the two degrees of freedom necessary to position the axis of the hole. Try to find some general principles that are helpful in deciding which motions should be

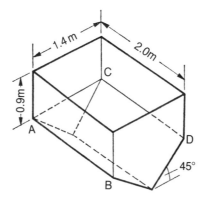

Figure 10.1   Hopper.

given to the work and which to the tool, and how the motions should be superposed, e.g. in a lathe, why put the cross-slide on top of the longitudinal slide and not the other way round?

Q.10.6. A silo for a military missile has a cover consisting of a horizontal slab of reinforced concrete 4.5 m square, weighing 200 tonnes. To fire the missile the slab must be moved 3 m to one side, and it is desirable to be able to cover and uncover the silo as quickly as possible. Analyse the problem and suggest suitable means. An electrical supply of 80 kW is available.

Q.10.7. Comment on the following passage from p. 36 of *Steam Power in the 18th Century* by Cardwell, which is about Watt's speculations on the poor performance of a miniature Newcomen engine.

> Now, Watt knew that since steam is condensed by cold metal surfaces, the greater the surface exposed the more the amount of steam condensed. Imagine a cylinder, open at one end, and of 10 in radius and 10 in long; also a 'model' of it built to a scale of 1 in 10 and therefore 1 in radius and 1 in long. Then to every square inch of surface of the big cylinder there will be $3\frac{1}{3}$ in^3 of volume; to every square inch of surface of the model there will be only $\frac{1}{3}$ in^3 of volume. A cubic inch of steam in the model is, therefore, exposed to much more cold surface than is the same amount in the big cylinder. Relatively more will be condensed and relatively more 'wasted'.

Q.10.8(2). A shaft is to be locked against rotation in one direction by a ratchet wheel and pawl mechanism. The torque to be resisted (T) is large and the maximum acceptable angle (θ) between successive locked positions is small. Show that there is a lower limit on the size of the ratchet wheel (i.e. the volume of the pitch cylinder) and comment.

Q.10.9(2). A gimbal ring with a mean diameter of 1 m is subject to loads normal to its plane at four equally spaced points, N, E, S, W. All the loads are of 100 kN, but those at N and S are oppositely directed to those at E and W. Estimate the maximum bending moment in the ring, and any other quantity relevant to a choice of cross-section. No heavy calculations are necessary, but symmetry considerations help.

Q.10.10(1). A meal is to consist of items X, Y, and Z and to contain at least 180 units of protein and 300 units of carbohydrate. What is the lowest-priced meal which meets these requirements?

	Units per penny	
	Protein	*Carbohydrate*
X	5	4
Y	2	8
Z	6	3

Q.10.11(2). Figure 10.2 shows a portion of the tower of a suspension bridge looking north along the roadway. Due to a westerly wind loading, the tower in the region shown is subject to a shear force S which may be regarded as constant over the bay. Estimate the volume of steel required in *one* cross-bar, using the maximum shear stress (Tresca) criterion with allowable *tensile* stress, f. Compare this with the volume required for diagonal braces AC and BD, and find the ratio for $p = 30$ m, $b = 23$ m, $d = 11$ m.

Q.10.12. Figure 10.3 shows a mechanism for opening the doors of underground

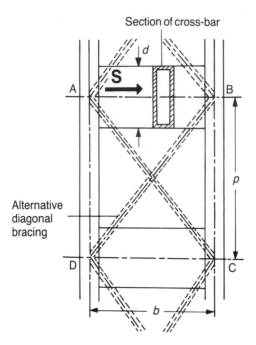

Figure 10.2   Suspension bridge tower.

trains. Why is it impracticable in the form shown, and what is the least alteration that will make it workable?Comment on the matching.

Q.10.13. Figure 10.4 is a diagram of a baling press that is to compress a loose fibrous material between a fixed platen P and a moving platen M by a single rotation of the crank C. Two arrangements are possible, (1) with the fixed platen at P_1 and (2) with the fixed platen at P_2. Find a few respects in which the schemes may be compared, and state which has the advantage in each case.

Q.10.14. It is necessary to ensure that a long rigid object, which is free to move a small distance horizontally and transversely to its length at each end, should remain parallel to a given direction, i.e. its degrees of freedom must be reduced from two to one, of translation perpendicular to its length in the horizontal plane. Sketch

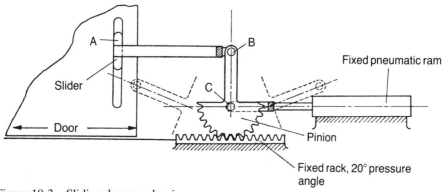

Figure 10.3   Sliding door mechanism.

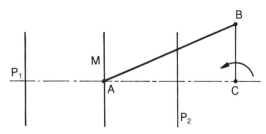

Figure 10.4   Baling press.

kinematic devices to apply this kind of constraint. What figures of merit could you apply to your solution?

Q.10.15(3). A ship may be regarded for the purposes of this question as a rectangular steel tube filled with a solid cargo which makes the combined specific gravity 0.50 (relative to seawater), of which 0.42 is due to cargo. The depth of the hull is 20 m, and the centre of gravity of the ship and cargo is 10 m above the bottom.

(a) find the beam of the ship if the metacentric height is 1 metre,
(b) because it is the cheapest way of making a ship bigger, it is proposed to make it deeper, keeping the metacentric height at 1 m by fitting fixed ballast of specific gravity 7.5 relative to seawater in the ship's bottom. As a first step in the economic study, find by how many tonnes the ship's displacement will be increased for every tonne of extra cargo. If you can, find a treatment which keeps the algebra simple.

Q.10.16(3).   A suction hovercraft travelling on a ceiling presents an almost insoluble control problem because it is inherently unstable—if it moves a little further away from the ceiling, the pressure in the lift pad will rise, and the vehicle will fall unless the exhauster can be stepped up in power very quickly. Yet at one time a suction hovertrain was being studied in France.

Enunciate the geometric property of the lift pad that is essential for stability in this respect in both 'suck' and 'blow' hovercraft, and find a configuration for 'suck' hovercraft which has this property.

Q.10.17. An old puzzle consists of two square-section pieces of wood joined end to end with each face of the joint showing a dovetail (Figure 10.5). It is possible to

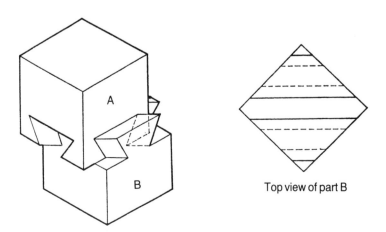

Figure 10.5   Dovetail puzzle.

make a similar puzzle consisting of two pieces of wood of equilateral triangle section which are joined together, with no voids in the interior, so as to show a dovetail on every face. How?

Q.10.18(3). A man climbs up out of a ravine mouth whose form is given by $z = x^2 e^y$, where z is the height and the units are hundreds of metres. He starts at the origin of co-ordinates and climbs always up the line of greatest slope to a plateau at $z = 4$. What is the width of the ravine where he emerges?

Q.10.19. Make a combinative study of ways of closing the end of a cylindrical vessel subject to internal pressure, and list the circumstances in which the various alternatives have advantages.

Q.10.20. To aid in manoeuvring them, some modern ships are pierced through from side to side by a horizontal tunnel containing a propeller. To drive such a propeller it is proposed to fit it with tip-jets ejecting water piped in through a hollow hub from the ship's fire-and-bilge pumps, which can pump seawater at approximately 0.7 N mm^{-2}. Comment on the matching problem.

Answers

A.10.1 If ω is the angular velocity of the propellers so that $20\,\omega$ is the gas-turbine speed, the interconnecting shafts should run at some intermediate speed, say, $k\omega$. If k is large, the torque will be small and the shafts light, except that the critical-speed problems will limit this effect.

If we put the engines in line with the propellers, the maximum power to be handled by the interconnecting shafts is $\frac{3}{4}E$ (where E is the power of one engine) against E if the engines are elsewhere. The difference is greater than appears, in that the $\frac{3}{4}E$ is an emergency power while the E is a normal service requirement, and so there is a strong argument for putting the engines in line with the propellers. Can we find an argument for putting them elsewhere?

The engines in such an aircraft will be a sizeable fraction of the weight of the fuselage. Putting the engines in pairs at the wing tips will reduce the bending moment in the wings over most of their length, reducing the structural task, perhaps by one quarter compared with the case with engines directly behind the propellers.

Notice that with the engines in line with the propellers, the central interconnecting shaft need only be rated at $\frac{1}{2}E$. Some complicated arrangements can be made to take advantage of this, but the weight saving is likely to be very small, as it is with schemes using multiple meshings of bevel gears. The best arrangement seems to be:

(a) multiple pinion simple epicyclics dropping engine speed to about $7.5\,\omega$,
(b) bevel drives of about 1:1 ratio interconnecting the $7.5\,\omega$ shafts,
(c) further simple epicyclics of 7.5 reduction ratio.

A.10.2. Yes, as part of the capital cost.

A.10.3. 2.5 tonnes of seawater yield 1 tonne of fresh water, and the work done in pumping to $15 \times 10^6 \text{ Nm}^{-2}$ is

$$1.25 \times 15 \times 10^6 \times 2.5 = 46.8 \times 10^6 \text{ J}$$

Of this, $0.8 \times 15 \times 10^6 \times 1.5 = 18 \times 10^6$ J is recovered in the turbine, leaving an energy consumption of

$$28.8 \times 10^6 \text{ J or 8 kWh per tonne of fresh water.}$$

Of this, the unavoidable work is that of forcing the fresh water through the membrane, or

$$1.25 \times 15 \times 10^6 \times 1 = 18.8 \times 10^6 \text{ J}$$

The remainder is due to pumping the other 1.5 tonnes into the high-pressure region and passing it out again through the turbine, and can be avoided simply by admitted the seawater and removing the brine at atmospheric pressure, i.e. by using a batch process, filling the plant at atmospheric pressure, pressurising until the fresh water expelled is 0.40 of the original charge, and then depressurising before refilling with seawater. There would be some power needed to compress the seawater and stretch the plant, which would not be negligible, and the fatigue condition on the plant would be expensive. The pressure over the membrane is roughly proportional to the salt concentration in the brine—you might try working out the optimum brine:fresh-water ratio k for a machine efficiency $\eta \{k = 1/\sqrt{(1 - \eta^2)}\}$.

A.10.4.   (a) The writer took 2 mm sheet for the lining of the two-tier version. At the depth of 1.5 m, the 'hydrostatic' pressure is

$$p = 1.5 \times 9.81 \times 600 \text{ N m}^{-2} = 0.0088 \text{ N mm}^{-2}$$

giving a bending moment in a span d encastered at each end of $\frac{1}{12}pd^2$, and this must be equal to or less than $fz = 120 \times 2^2 \text{N}/6$.

Hence the allowable $d = 330$ mm, and from this may be deduced suitable pitchings for the reinforcements (which above the doors can with advantage be continuous horizontal 'rings'). The total weight of steel was about 300 kg. For a three-tier structure, the thinnest likely sheet was taken as 0.6 mm. With a flat sheet on the inside and a sheet with rectangular corrugations 20 mm deep on the outside spot-welded to it, reinforcements can be confined to the top, the level of the door hinges and the meeting edges of the doors. The weight of steel needed was estimated to be 210 kg. The material saving of perhaps £50 per hopper would almost certainly be lost in additional fabrication and finishing costs, and the two-tier design is probably cheapest, apart from the vulnerability of the lighter structure to corrosion.

(b) Springs or deadweights, with a retaining catch or some delay to ensure complete emptying.

If the doors are latched at one end only, the torque in a door can be calculated to be 9600 Nm. This can be carried out to the latch end by a tube 80 mm in outside diameter and 5 mm wall thickness, which will still have sufficient reserve of strength to act as a reinforcement in bending. The extra cost is likely to be less than £20 per hopper, which would probably be saved on the latching arrangements.

Another approach is to put two reinforcing arms on the door, say 600 mm in from the ends, and fit the single latch on the end of one of these arms. A smaller tube will then serve to take torque and bending moment (from the bending moment alone the ideal position of the arms is 410 mm from the ends—check this).

It may be worth considering making a door extend, say, three-quarters of the way across the plan view of the hopper, and replacing the other by a fixed slope which will now be inclined at 63.4° to the horizontal.

A.10.5. Important criteria are overall size, stiffness and the effects on accuracy. In some cases, the nesting order is dictated, e.g. generating movements must be 'inside' or 'farther from earth than' positioning movements. In deciding whether to move work or tool, it is usually better to move the tool for concave surfaces and the work for convex surfaces. For flat surfaces move the tool if the work is very large. It is usually better to nest motions of smaller travel inside those of larger travel. All these are guides only—each case must be treated on its merits.

A.10.6. Suppose 70 kW is available at the motor shaft and this is applied by a frictionless infinitely variable drive for the first 2.5 m. Then the kinetic energy is $70000t$ J, where t is the elapsed time in seconds and so

$$70\,000t = \tfrac{1}{2} \times 200 \times 10^3 v^2, \; v = \frac{\mathrm{d}x}{\mathrm{d}t} = 0.84t^{1/2}$$

where v is the speed in metres per second, x is the travel in metres. Hence

$$x = 0.56t^{3/2}, x = 2.5 \text{ when } t = 2.7 \text{ and } v = 1.38$$

If the 'eyelid' travels the last 0.5 m at an average speed of $\tfrac{1}{2} \times 1.38$ m/s under a uniform deceleration due to some buffer, the total time is

$$2.7 + \frac{0.5}{0.69} = 3.4\text{s (say)}$$

(We cannot quite reach this figure because we cannot start with an infinite acceleration.)

The problem can be eased by making the slab the mass of a spring-mass or pendulum-type vibratory system, with its 'at rest' position half-way between open and shut. If the natural frequency is 0.5 cycles per second, the opening and shutting times will be one second. To open the shut 'eyelid' we have only to release the catch which holds it shut. It automatically starts to vibrate, but is caught as soon as it comes to rest for the first time in the open position.

A steel coil spring working at a shear stress of 500 N mm^{-2} stores strain energy at a maximum density of 1.63×10^{-3} J mm^{-3} or 210 J kg^{-1} (Section 4.5) with an efficiency of utilisation of 0.5, which brings the average energy stored down to, say, 100 J kg^{-1}. The energy of a vibratory system is $\tfrac{1}{2}ma^2\omega^2$, where m is the inertia, a the amplitude and ω the angular natural frequency. Here $m = 200$ tonnes, $a = 1.5$ m and $\omega = \pi s^{-1}$, so the energy is 2.2×10^6 J, requiring springs weighing about 22 tonnes, which is rather discouraging. Air cylinders would be practicable, however. Mounting the slab on 'switch-back' rails would probably defeat its protective function.

In any 'vibratory' solution, the decay of amplitude could be overcome by biasing the 'position of rest' away from the current position of the slab, i.e. by increasing the pressure in the cylinders on the 'compressed' side. The size of compressor required would depend on the frequency of operation demanded.

A.10.7 More simply stated, the argument is that the mass of steam in the cylinder is proportional to L^3, while the area on which steam is condensing is proportional to L^2. This argument ignores the time factor—by the reasoning of Section 4.7 the most suitable simple assumption is that velocities remain constant, so that the duration of the stroke is proportional to L and the product of the condensing area and the time it is exposed is proportional to L^3 and therefore the volume. In this case the little engine would be no worse off than the big one.

In more detail, the heat-transfer problem involves (a) the 'resistance' to heat

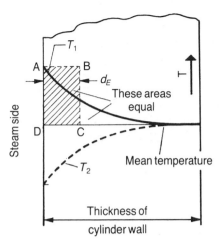

Figure 10.6   Temperature distribution in cylinder wall.

transfer at the fluid/solid boundary and (b) the unsteady conduction in the wall.
Three simple special cases can be distinguished, (1) where (a) is the bottleneck and
the wall itself is to all intents and purposes an infinite heat sink, (2) where (a) is
negligible so the inside face of the wall is always at steam temperature, and (3)
where both (a) and (b) are negligible and the wall heats and cools right through on
every stroke. Case (1) is the one we have just considered, and case (3) also puts the
big and little engines on an equal footing. In case (2), however, the temperature (T)
distribution through the wall will be somewhat as in Figure 10.6, where T_1 is the
temperature at the end of the inlet stroke and T_2 is the temperature at the end of the
condensing stroke. The shaded rectangle ABCD has the same area as that under
the T_1 curve, so that d_E is the *effective thickness* of the cylinder wall in this context:
d_E depends on the conductivity, specific heat, and density of the material of the
wall, and is also proportional to the square root of the duration of the stroke, and
hence to $L^{1/2}$. Case (2) reduces to case (3) with the cylinder wall thickness reduced
to d_E, the cooling effect is proportional to the *effective* cylinder water equivalent
and hence to $L^{5/2}$, and since $\frac{5}{2}$ is less than 3, the big engine is better off than the little
one: d_E is another useful concept.

$$d_E = \left(\frac{4kt}{\pi c \rho} \right)^{1/2}$$

where k is the thermal conductivity, c is the specific heat, ρ the density of the wall
material and t is the duration of a stroke.

A.10.8. If we fix the pitch radius r of the ratchet wheel, the pitch of the teeth is θr
and this fixes the size of the teeth and hence the load per unit facewidth they can
carry at, say $A\theta r$, where A is constant. Now the load on the pawl, if we use only one,
is T/r so that the facewidth is

$$\frac{T}{A\theta r^2}$$

and the pitch cylinder volume is

$$\frac{T}{A\theta r^2} \times \pi r^2 = \frac{\pi T}{A\theta}, \text{which is fixed.}$$

If we use 2N pawls, the pitch cylinder volume is reduced to $\pi T/2NA\theta$. The best way will usually be to fit the pawls in diametrically opposed pairs (to balance the forces on the wheel), to increase the tooth pitch to $N\theta r$, and to stagger the N pairs of pawls so that successive pairs engage at angular intervals θ.

When θ is small enough, a friction freewheel which makes θ 'infinitely' small will be smaller than a ratchet and pawl mechanism. The sprag type is most compact and an interesting disposition problem. The analysis is similar to that of Section 4.9.

A.10.9. The ring cross-section is subject to shear force, bending moment and also possibly torsion, and we need to know the maximum values of these. The maximum shear force is ±50 kN and obtains everywhere except under a load. The bending moment must be zero at NE, SE, SW, NW for reasons of symmetry. In the upper part of Figure 10.7 the E and S quadrants have been shown with bending moments at the ends, and it is clear that if the sign of the moments is correct in one it must be wrong in the other—the only possibility is that the moments are zero. The maximum bending moments must occur under the loads, and can be found by the equilibrium of the eastern half ring (Figure 10.7)—the maximum M is 25 kN m. The maximum torque occurs at NE, SE, etc. and can be found from the equilibrium of the eastern quadrant (Figure 10.7) by taking moments about the NE-SE axis, giving

$$2/\sqrt{(2)}T = 100 \times 0.5\left(1 - \frac{1}{\sqrt{2}}\right) \text{kN m}, \ T = 10.4 \text{ kN m}.$$

Feilden records a case [37] of this torque being overlooked.

A.10.10. 30p worth of y, 20p worth of z

If x, y, z pence are spent on X, Y, Z, the condition that there is just enough protein is

$$5x + 2y + 6z = 180$$

which is a plane dividing the *feasible* from the *non-feasible* region of x, y, z space. Similarly, from the carbohydrate requirement,

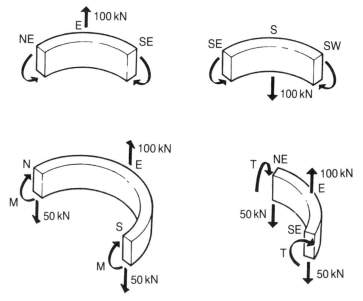

Figure 10.7   Bending and torsion in gimbal ring.

$$4x + 8y + 3z = 300$$

One way of finding the answer is to draw the traces of these planes in the xy, yz, zx planes, and also the traces of planes of constant cost ($x + y + z = C_T$).

A.10.11. There will be zero bending moment in the legs halfway between crossbars and the shear in each leg is $S/2$. The bending moment in the ends of the crossbars is $\frac{1}{2}pS$, and the shear force in them is pS/b. The combined area of the flanges must be

$$2 \times \frac{\frac{1}{2}pS}{fd} \text{ and that of the shear webs } \frac{2pS}{bf}$$

The total volume of material in one crossbar is thus

$$\frac{Sp}{f} \left[\frac{b}{d} + 2 \right]$$

A pair of diagonal braces will be of length $(b^2 + p^2)^{1/2}$ and carry a combined load

$$S \, \frac{(b^2 + p^2)^{1/2}}{b}$$

The volume of material required is thus

$$\frac{S}{f} \cdot \frac{b^2 + p^2}{b}$$

so the ratio of volumes is

$$\frac{bp \left(\dfrac{b}{d} + 2 \right)}{b^2 + p^2} = 1.98$$

for the given dimensions. It is possible to graduate the thickness of plating in the flanges of the crossbars and reduce this ratio. However, the crossbars also induce bending moments in the legs which have to be paid for as well and add to the functions competing for a share of the allowable stress.

Notice that even this structurally uneconomical form requires that the crossbars must be more than half the depth of the clear space between and are much the most heavy-looking part—quite out of scale with the slender grace of the rest of the bridge and performing, for all their ponderousness, a relatively small task.

A.10.12. To avoid jamming at A the slider would have to be either very long or frictionless: the first is awkward to accommodate and the second expensive. A pivot rather than a slider at A solves the problem. Also, some sort of crosshead at C seems desirable to take the side thrust from the gear teeth.

Since it provides an increase of mechanical advantage at both ends of the travel, where it may be required, the matching seems good.

A.10.13. Compactness: scheme 2 is clearly more compact.

Complexity: the connecting rod and crank must be duplicated in scheme 2.

Structural task, static parts: the static parts carrying the pressing force have perhaps half the structural task in scheme 2.

Matching: the pressing action can occupy a greater fraction of a turn of the crank in 2, giving a smaller maximum torque.

A.10.14. Figure 10.8 shows three possibilities. Of these (a) uses material most economically and (b) is the tidiest. Little can be said in favour of (c), but if the inner

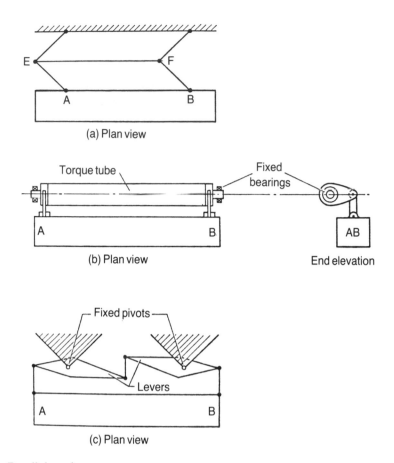

Figure 10.8   Parallel motions.

arms of the two similar levers are made much longer than the outer ones reasonable structural economy is possible. (Why is this the best proportion for the levers?—see A.5.11.)

In this sort of problem the movement to be accommodated is generally small, but on the other hand the parallelism required must be correspondingly good. With large loads to be moved, high stiffness is usually required, and here the rather clumsy version (a) wins. As a figure of merit, we could use

$$\frac{\text{angular stiffness}, k \,(\text{say}, \text{in Nm/rad})}{V, \text{volume of material used for linkage}}$$

It is interesting that if we try to make this non-dimensional we do not obtain a logical form, since k/EV is non-dimensional but does not involve the length AB (Figure 10.7) over which parallelism is to be maintained, which makes nonsense of it. The reason for this anomaly is that the structural task (Section 8.1) is nil. In the mechanism of Figure 10.8(a), for example, all the members can be vanishingly small except for EF. Now by the argument of A.5.11, EF can also be made as thin as we wish while keeping its effective stiffness, simply by gearing its motion up—the function it performs is analogous with that of the brake cable. While this is not practical, it does show that there is no *absolute* lower bound for V for a given k.

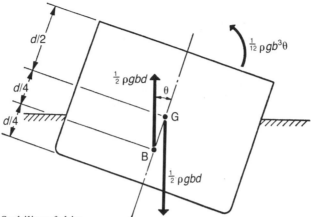

Figure 10.9   Stability of ship.

A.10.15. (a) If b is the beam, d the depth, ρ the density of seawater, then per unit length of ship the displacement is $\frac{1}{2}\rho bd$. The centre of buoyancy B is $d/4$ below the centre of gravity G, so that when the ship is tilted through angle θ, there is an overturning moment per unit length of $\frac{1}{8}\rho gbd^2\theta$ (Figure 10.9). The restoring moment of the water surface is $(1/12)\rho gb^3\theta$, so that

$$\tfrac{1}{12}\rho gb^3\theta - \tfrac{1}{8}\rho gbd^2\theta = \text{displacement} \times \text{metacentric height} \times \theta$$
$$= \tfrac{1}{2}\rho gbd\theta \times \text{metacentric height}.$$

(In future equations of this form we shall leave out the factor $\rho gb\theta$.) Putting $d = 20$ m, metacentric height $= 1$ m, we find $b = 26.8$ m.

(b) 2.7. This is a differential study, but it is difficult by differentiation. Use actual figures, and start by assuming a given *increase in draught* to make the algebra simple. If the draught is 11 m, and the centre of gravity is now x m above the bottom, we can find x directly from the condition that the metacentric height is 1 m.

$$\tfrac{1}{12}b^2 - 11[x - 5.5] = 11 \times 1, x = 9.95\,\text{m}$$

We now use the idea of the ballast as coexisting with cargo and steel but having just an *incremental* specific gravity of 7. Then the draught

$$= 7y + 0.5d = 11 \tag{1}$$

Also, taking moments about the bottom, where y is the depth of ballast in metres, the mass moment is

$$11x = 7\frac{y^2}{2} + 0.5 \times 0.5d^2 \tag{2}$$

From equation 2, ignoring the y^2 term, $d = 20.95$ m. From equation 1, $y = 0.075$ and substitution in equation 2 gives a negligible change in d. (This is much better than substituting and solving quadratics.) Hence answer—note additional draught due to cargo $\simeq 0.42(0.95 - 0.075)$ m.

Now as the displacement increases by nearly three tonnes for every tonne of additional cargo, the increase of propulsion costs (which are about one-third of the whole—Section 3.3) is nearly equal to the costs of transporting one tonne in the original ship. When we consider also the cost of the ballast (which might be pig-iron) the case against the 'deep ship' looks complete. If we use concrete at perhaps one-tenth of the price, the ratio of increase of displacement to increase of cargo becomes even more unfavourable. On the other hand, we have not taken account of the reduced thickness of deck and bottom plating possible because of the increased d. We can conclude that the pig-iron-ballasted ship is not viable, but the concrete-ballasted one may be a 'runner' and warrants a more careful study, taking account of double bottoms, the ends of the ship, etc.

A.10.16. Round more than half the periphery of the lifting pad the lifting pad side of the gap between pad and support must lie above the support. Imagine a suction hovertrain running on the bottom flange of an I-beam, using a rectangular lifting pad which curls over the flange at the sides so that the gap along the sides has the pad above the support (Figure 10.10). Then if the pad is substantially longer than it is wide, the condition for stability will be met.

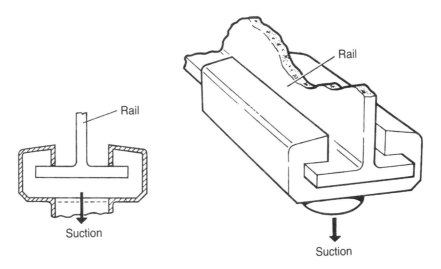

Figure 10.10   Pad configuration for stable suction hovercraft.

A.10.17. There are two solutions at least. Section 2.3 gives the key—one solution is to make the joint a turning pair with the centre line of one side as axis. The other solution is a combined sliding-convergent joint, and is rather less obvious. There may well be other solutions the writer has not found.

A.10.18. Slightly over 300 m. The contours are given by $z = $ const. $= x^2 e^y$ Differentiating with respect to x for constant z gives

$$2xe^y + x^2 e^y \frac{dy}{dx} = 0, \text{ or } \frac{dy}{dx} = -\frac{2}{x}$$

Now on a contour map with y, x axes on it, the slope of the contours is $-2/x$ so that the slope of the line of steepest slope, which is at right angles to the contours, is $x/2$, and the differential equation of the man's path on the map is $dy/dx = x/2$, or $y = x^2/4$ since he starts at the origin. He arrives at the plateau at the point

$$z = 4 = x^2 e^y = 4ye^y$$

Solving (by successive substitutions, say) $y = 0.568$ (roughly) $x = 1.5+$.

A.10.19. Useful classifying principles are:

(a) extra parts or not,
(b) number of surfaces in shear (clearly, there must be at least two).

Cases of no extra parts, besides welded, brazed and simple screwed joints, are the interrupted, stepped, screw threads (Welin screw) used for the breaches of big guns and the transversely sliding breach blocks of much artillery. Then there is the Welsh plug, a dished disc of material expanded by flattening into a circumferential groove in the bore of the cylinder.

   Important cases of two surfaces in shear are those where a plug is retained in the bore of the cylinder by a ring in a circumferential groove or against a shoulder. The ring may be got past such a shoulder by distorting it into a ellipse and slipping it through at an angle, or by making it in sectors—the former method has advantages with respect to sealing [52].

A.10.20. The characteristic or 'spouting' velocity c of the water from the fire- and bilge-pumps is roughly $(2 \times 700)^{1/2} = 37$ m/s, while for the propeller we seek about one-tenth of this or less (Section 5.2). Since the effective blade speed u will be nearly the same for both fluids, we seek a ratio of the u/c ratios (Figure 5.2) of about ten, and this is possible. But generally speaking the power is not adequate for the purpose, so this interesting combination of functions is not practical.

References

1 EUSEC conference, Copenhagen, September 1966.
2 Ghiselin, B. *The Creative Process*. New American Library, New York, 1952.
3 Ruhemann, M. *The Separation of Gases*. Oxford University Press, 1949.
4 'Moulton makes a better bicycle . . . again.' *Engineering* **233**, no. 7, p. 547, 1983.
5 Jung, I. 'Swedish Marine Turbine and Gear Development.' Society of Naval Architects and Marine Engineers, Spring Meeting, 1966.
6 Wankel, F. *Rotary Piston Machines*. Iliffe, London, 1965.
7 'Design leaders impress the market.' *Engineering*, **233**, no. 7, p. 547, 1983.
8 Alexander, C. 'Notes on the synthesis of form.' PhD thesis, Harvard, 1962.
9 Cardwell, D. S. L. *Steam Power in the 18th Century*. Sheed and Ward, London, 1963.
10 'Stirling Machine Developments.' *The Engineer*, Nov. 26, pp. 906–12, 1965.
11 Kays, W. and London, A. L. *Compact Heat Exchangers*. McGraw-Hill, New York, 1964.
12 Battersby, A. *Mathematics in Management*. Pelican Books, London, 1966.
13 Gass, S. *Linear Programming: Methods and Applications*. McGraw-Hill, New York, 1975.
14 Fox, C. *An Introduction to the Calculus of Variations*. Oxford University Press, 1950.
15 Dickinson, H. W. *A Short History of the Steam Engine*, p. 118. Cambridge, 1939.
16 Gifford, E. W. H. 'The Development of Long-Span Prestressed-Concrete Bridges.' *Structural Engineer*, p. 325, October, 1962.

17 Cox, H. L. *The Design of Structures of Least Weight*. Pergamon, Oxford, 1965.
18 MacMillan, R. H. and Davies, P. B. 'Analytical Study of Systems for Power Transmission.' *Journal of Mechanical Engineering Science*, **40**, 1965.
19 Smeed, R. J. 'A Theoretical Model of Commuter Traffic in Towns,' *Journal of the Institute of Maths Application*, **1**, 208–25.
20 Davies, W. J. 'Performance and Reliability of High-Duty Gearing for Aircraft.' Paper 16, Proceedings of the International Conference on Gearing, Institution of Mechanical Engineers, London, 1958.
21 Hausen, H. 'Verlustfreie Zerlegung von Gasgemischen durch umkehrbare Rectifikation.' *Zeitschrifte fur technische Physicen*, **6**, 272–7, 1932.
22 Ljungström, F. 'The development of the Ljungström steam turbine,' *Proceedings of the Institution of Mechanical Engineers*, **160**, 216, 1949.
23 Stodola, A. *Steam and Gas Turbine*. Smith, New York, 1945.
24 Payne-Gallwey, R. *Projectile-throwing engines of the ancients*. E. P. Publishing, Wakefield, 1973.
25 Homer *Odyssey*, book 21.
26 French, M. J. 'Force and corner-power relationships in two-regime and many-degree-of-freedom mechanisms.' *Mechanisms*, Institute of Mechanical Engineers, 1972.
27 Grylls, S. J. 'The History of a Dimension.' *Proceedings of the Institute of Mechanical Engineers*, **178**, part 2a, p. 1, 1963.
28 Tuplin, W. A. *Gear Load Capacity*. Pitman, London, 1961.
29 French, M. J. 'Gear Conformity and Load Capacity.' *Proceedings of the Institute of Mechanical Engineers*, **180**, part 1, 1965–6.
30 Glegg, G. J. *The Design of Design*. Cambridge University Press, 1969.
31 Bitter, F. 'The design of powerful electromagnets,' *Review of Scientific Instruments*, **7**, 1936.
32 Laithwaite, E. R. *Induction Machines for Special Purposes*. London, 1966.
33 Parkes, L. R. 'Design for a Marine Diesel Engine.' *British Welding Research Association Bulletin*, vol. 6, no. 4, 1965.
34 Niemann, G. *Maschinenelemente*, vol. 2. Springer, Berlin, 1960.
35 Borovich, L. S. *Lowering the weight of reduction units by a rational choice of basic parameters in increasing the loading of gearing and decreasing its weight*, ed. M. M. Saverin. Pergamon, Oxford, 1961.
36 Joughin, J. H. 'Naval Gearing—Wartime Experience and Present Developments.' *Proceedings of the Institution of Mechanical Engineers*, **164**, 1951.
37 Feilden, G. B. R. 'A Critical Approach to Design in Mechanical Engineering.' *Bulleid Memorial Lectures*, University of Nottingham, 1959.
38 Tuplin, W. A. 'Improved Mousetrap.' *The Engineer*, April 30, p. 773, 1965.
39 Osborne, A. F. *Applied Imagination*. Scribners, New York, 1963.
40 Dixon, J. R. *Design Engineering—Inventiveness, Analysis and Decision*. McGraw-Hill, New York, 1966.
41 Atkins, A. J. and Ellis, F. E. 'An Automatic Transmission for Small Cars.' *Proceedings of the Institution of Mechanical Engineers*, **181**, part 2a, no. 6, 1966–7.
42 ICI *Assessing Projects*. Methuen, London, 1970.
43 Pahl, G. and Beitz, W. *Engineering Design*. Design Council, London, 1984.
44 Matousek, R. *Engineering Design*. Blackie, Glasgow, 1963.
45 Rodenacker, W. *Methodisches Konstruieren*. Springer, Berlin, 1970.

46 Marples, D. L. *The Decisions of Engineering Design*. Institute of Engineering Designers, London, 1960.

47 Nervi, P. L. *Aesthetics and Technology in Building*. Harvard University Press, Massachusetts, 1966.
     All quotations from *Aesthetics and Technology in Building* are reproduced by kind permission of the publisher.

48 VDI 2222, Konzipieren Technischer Produckte, VDI Verlag, Düsseldorf, 1977.

49 Stubbs, P. W. R. 'The Development of a Perbury Traction Transmission for Motor Car Applications.' *American Society of Mechanical Engineers*, 80-C2/ DET 59, 1980.

50 Ehrlenspiel, K., Herstellkosten von Zahnrädern, VDI-Berichte, no. 488, pp. 97–105, 1983.

51 Booker, P. J. *Principles and Precepts in Engineering Design*. Institute of Engineering Designers, London, September 1962.

52 Anderson, H. H. 'Self-seal joint for very high pressures.' ASME paper no. 5, Pressure Vessels and Piping Division, Winter meeting, 1967.

Index

abstraction, level of 5, 8, 10, 102, 107
adhesives *199*
aesthetics 12
aircraft
   man-powered 104, 105
   undercarriage 201
   v.t.o.l. 42, 205–6, 210
   wings 139
alternatives, examining 192–3
alternators 136–7
   turbo 111–12, 114
Anglepoise lamp 127–8, 130–1
arbitrary decisions 187–9, 200, 201, 202
assets, using existing 170–1, 179–80, 193–4
autonomous local optimising principle
       (ALOP) 69, 86
auxiliary function 7, 43
availability 90

baby-carriage problem 45, 48
baling press problem 208–9, 215
ball joint problem 148, 149–50
balloon design 171
bathroom scales 100
beams 10
   fixed end 78, 170
   ideal concept 92
   optimisation problem 73, 75
   structural task 91–2
bearings, load balancing 155, 165–6
   roller 98–9, 169–70
bed, modelling example 189–92
bending moments *see* beams
bicycle, Moulton 8–9
binoculars 92
blade, suspension 106
blade attachment, rotor 15–17, 132–3,
     187–8
   see also rotors; turbines, gas
bolts, high strength friction-grip 45–7
bows, examples of matching 121–2
box girders 54, 78–80, 139
brachistochrone problem 65–6, 69, 86
brainstorming 196
brake caliper, disc 63, 134–6, 155
   see also handbrake

bridges 16, 106, 139
   see also suspension bridges

calculations, rough 76–85, 203
calculus of variations 66
candle, functions of 43
canoe, mainsheet problem 128–9, 131
car design 3
   body 11–12
   doors 197
   engine position 13–14, 28–30, 32
   handbrake 123–6, 129, 131
   suspension 140–1, 201–2
   transmission 93–4, 200–1
   see also engines, internal combustion
centring devices 133–4, 167–9
checklists 180, 203–5
chess example 5–6
clamp, diaphragm 77–8
clarity of function 8
classification
   optimisation 62–6
   options 14, 16, 25–7
      car engine 45, 47
clutch, motor-mower 45, 47
coal mining problem 64–5
coinage problem 201
combination of functions 24
combinative ideas 13–48, 203
   car design 13–14, 28–30, 32
   description 13
   hybrid solutions 33–4, 42, 198
   pressure vessels 210, 219
   reduction gear 25–6
   Rodenacker method 25, 27–8, 29
   rotor example 15–22, 31
   tanker example 23–5, 31, 33–4
   v.t.o.l. aircraft 205–6, 210
   Wankel engine 25
   wave energy converters 34–9
compliant member 161
components, design of 4, 11, 12, 198
   standard 185
compressors, axial flow 15–22
   see also blade attachment, rotor;
     rotors

computers   7, 64
    as components   12
    for rough calculations   77
    in detailing   3, 10
    in modelling   11–12, 196
conceptual design, meaning   1, 3
connecting-rod problem   148–9, 150–1
cookers, electric   194
corner power   10, 121
    description of   93, 94
corner work   121, 123–5, 135
cost of performance figures   57–9
costs   1, 178–86, 203
    annual charge basis   49
    design charges   205
    gearing   180–4
    heat exchangers   55–7, 73–4, 75
    interest charges   49–50
    minimum   2–3, 181, 207, 214–5
        see also   optimisation
    pressure vessels   184–5
    reducing   181–3
    refrigerators   59
    suspension bridges   53–4
    tankers   50–2, 63
crane, level-luffing   127, 130
    mobile   129, 131
cryogenic plant   126, 129–30

daisy-wheel typewriter   44
degree of freedom, single   50–4, 63
degrees of freedom
    choice   187–9, 200, 201
    matching   113–14, 208–9, 215–6
    multiple   55–7, 63–6, 206–7, 212
    reduction gear   126
    spatial   153–4, 200, 201
desalination plant   206, 210–11
design
    costs   205
    meaning of   1
    philosophies   8, 38–9
    process   1–4
design methods, limitations   4
detailing   2, 3, 10
diaphragm clamp   77–8
differential studies   31, 59, 83–5
dimensional analysis   98–9
disc brake calipers   63, 134–6, 155
    see also   handbrake
discounted cash flow   50
discs, turbine   80–3
disposition   132–52, 203
    alternator example   136–7
    disc brake example   134–6
    gear teeth design   142–5
    rotor blades   132–3
    rotor shaft   133–4
    structural form   139–42
    suspension bridge   146

diversification of approach   6
doors
    car   197
    underground train   207–8, 215
dovetail puzzle   209–10, 218
drag of ship   50–2
drawings, working   2, 3, 10
duct hinge problem   172, 175

economy, structural   170–1
efficiency of use   88
elastic design   157, 160–6, 198–9, 203
    below yield stress   69–71
    epicyclic gearing   161–3
    torsion rod   163–4
elastic devices, pros and cons   167
electrical machines   96
electrical power supply   73–4, 111–12, 126,
        129
electromagnets   146–7
engines
    diesel   171
    heat   149, 151–2, 192
    internal combustion   132, 198
        scale effects   94–5
        valve gear   45, 47
    Phillips Stirling-cycle   40–1
    Rolls-Royce
        Gazelle   158
        Gnome   20, 22
        Gyron Jnr   20–1
        Spey   18
    steam   10, 40, 82–3, 207
    Stirling-cycle   10, 40–1
    Wankel rotary   25

feasible area   61–2
feather, disposition example   137–9
Fibonacci search   63
figure of merit   87–9, 103, 105
finite element methods   11–12, 77
foil production control   25, 27 8, 29
force path   123
frames, pin-jointed   170–1
function
    checklist   204
    clarity of   8, 198
    optimum   28
    redistribution   40–3
functional analysis   14, 28, 197
    compressor rotor   15–17
    connecting-rod   148–9, 150–1
    LNG tanker   23
    tin-opener   45, 48
    wave energy converter   34

gearing
    automatic transmission   200, 201
    change of viewpoint   102

epicyclic
  common fallacy   201, 202
  elastic design   160–3
  flexible pin problem   172, 175–6
  kinematic design   155–60
  star gear   183–4
  for contra-rotating propeller   26, 183–4
  load sharing   167
  reduction
    contra-rotating propeller   26, 183–4
    cost reducing of   181–3
    performance measure   180–1
    Rolls-Royce Gazelle   158
    tanker turbine   25–6, 109
    twin layshaft   173, 177
  scale effects   97, 99–100
  tooth design
    involute   142–5, 148, 149
    non-involute   145
geometrical example   4
gimbal ring   207, 214
golf-ball typewriter   44
Goodman diagram   146
goodness factor (Laithwaite)   147

handbrake   123–6, 129, 131
heat exchangers   6
  matching   117–21, 127, 130
  optimisation   55–60, 73–4, 75
heat transfer   55–6, 60
helicopter engine mounting   154, 170
hill-climbing technique   64, 210, 218
holding-together power   10, 68, 179, 184–5
hoop strength, effective   72
hopper problem   206, 211
hotel bedroom problem   147–8, 149
hovertrain, suction   209, 218
hybrid designs   42, 198
  LNG tanker   33–4
  rotor shaft   20, 22

insight into problems   4, 5–6, 76–106, 203
  box girder design   78–80
  change of viewpoint   100–2, 105
  diaphragm clamp   77–8
  refrigerators   89–91, 100–2
  rotor shaft diameter   83–5, 103, 105
  scale effects   94–9
  springs   87–9
  turbine shroud   80–1
  use of ALOP   69, 86
  use of computers   12
  use of differential studies   59
instruments, design of   169
insulation of tankers   23–5
interaction
  first order   28, 29
  second order   28
  strong and weak   28

interest charges   49–50
inversion   135–6, 192–3

joint, flanged   188–9
joint design, pressure vessel   141–2
joint efficiency   87, 120

kernel table of options   32–3
kinematic design   153–9, 166–9, 198–9, 203
  disc brake   155
  epicyclic gearing   155–60
  helicopter engine mounting   154
  principle of   153
kinematic devices, pros and cons   166–7
kitchen cupboards problem   148, 149
knights (chess) problem   5–6

Lagrangian multipliers   64–5, 84
lathe problems   171–2, 174–5, 206–7, 212
letter balance problem   126, 127, 130
Linde-Fränkl process   7
linear programming   61, 62, 63
Ljungström's suspension blade   106
load diffusion   42, 67, 171
logical chains   135–6, 193–5, 199

machine tools   77–8, 206–7, 212
maintenance and inspection   204
manufacture, checklist   204
matching   73, 109–31, 203
  car handbrake   123–6
  cut-price solution   124–5
  degrees of freedom   113–14
  electrical power   111–12
  heat exchangers   117–21
  off-design-point   126, 129, 131
  ship propulsion   107–10, 112–13, 210, 219
  spring and task   115–17
mathematical techniques
  calculus of variations   66
  Goodman diagram   146
  hill-climbing   64, 210, 218
  Lagrangian multipliers   64–5, 84
  linear programming   61, 62, 63
  maxima and minima   62–4
  steepest descent   64, 210, 218
maxima and minima, finding   62–4
meal, minimum cost   207, 214–5
membrane tanks   24–5
models   196–7
  mathematical   189–91, 201–2
motorcars   see cars
Moulton bicycle   8–9

Nervi, P.L.   76–7, 80
Newcomen engine   10, 40, 207
nuclear reactors
  as insight example   97
  costs   185

duct hinges   169
holding-together power   184–5
in submarines   47
matching compressors   129, 131
matching heat exchange   117–121
r.c. pressure vessel   40, 41, 42

oil pipeline problem   74, 75
oil tankers   50–1, 63
optimisation   49–75, 114, 203
    ALOP   69
    classification of problems   62–6
    coal mining   64–5
    compressor shaft   83–6
    gas turbine example   57–9
    gearing   181
    heat exchangers   55–60, 73–4, 75
    minimum cost criterion   49–50, 51–2,
        53–4, 61–2
    multiple degrees of freedom   55–7, 63–6
    of shape or function   65–6, 69
    rotating discs   66–73
    ship design   209, 217–8
    ship speed   73, 74, 206, 210
    single degree of freedom   50–4, 63
    suspension bridge example   52–4
    tanker speed   50–2, 63
    turning value problems   62–4
    variable tolerances   60–2
    optimum function   28
    options
        evaluating   28–30
        kernel table of   32–3
        pivotal   32
        reduction of   17, 24, 28–32
        table of   see   table of options
        think of advantages   31

parameters, using   17, 65, 73, 91
parametric mapping   17–22, 47, 109–10,
    113
parts   see   components
Pelton wheel   42, 110, 114
philosophy, design   8
pivotal option   32
plastic design   71–3
polymers   140, 199
power supply
    electrical   73–4, 111–12, 126, 129
    in submarines   45, 47
preconceived ideas   196
pressure vessels
    combinative ideas   210, 219
    costs   184–5
    joint design   141–2
    matching problem   129, 131
    reinforced concrete   40, 41, 42
problem, analysis of   2
problem, statement of   2
programming, linear   61, 62, 63

proportion, designer's feel for   99–100
pumps
    swashplate hydraulic   74, 75
    water, steam driven   122

questions, ask stupid   199

radar   82
ratchet wheel problem   207, 213–14
ravine problem   210, 218
records, keeping design   4
reduction gearing   see   gearing, reduction
refrigerators   59, 89–91, 100–2, 120–1
    Phillips C70   41
repêchage   31, 43–4
rocket, ELDO   171
Rodenacker design method   25, 27–8, 29
roller bearings   98–9, 169–70
rotating discs   31, 66–73
    hub design   69–73
rotors
    choice of type   15–22
    disc   16, 17, 18–20, 45, 48
        centring devices   133–4, 167–9
    drum   16, 18
    optimum shaft diameter   83–5, 103, 105
    solid shaft   16, 18
    three-bearing shaft   172–3, 176
    virtual shaft   16–17, 19, 85–6
rough calculations   76–85, 203

safety   2–3, 205
scale effects   94–9, 207, 212–3
scale factors   55–6
schemes, embodiment of   2, 3
schemes, meaning of   1
seal, Phillips piston   40–2
sensitive heat   10
shape, matching   117
shape, optimising   65–6, 69
shear force   see   beams
shear stress criterion   66, 87, 163, 207
ship design   139, 209, 217–8
ship propulsion   107–10, 112–13, 183–4,
    210, 219
silo, missile   207, 212
simulation   11   see also   models
slenderness ratio   17
solar panel   140–1
space limitation   see   disposition
spatial degrees of freedom   153–4, 200, 201
specific speed   113
springs
    cantilevered   165, 166
    compression   87–8, 115–17, 127–8,
        130–1, 140–1
    fluid   104, 105
    leaf   88–9
    watch   104, 105
standard components   185

steepest descent method   64, 210, 218
step size reduction   7
steps, inventive   7
Stirling-cycle engines   10, 40–1
strain energy in spring   87
stress analysis   11–12, 77
structural form   139–42
submarine, magneto-hydrodynamic   102–3, 104–5
submarine, power supply to   45, 47
suspension bridges   106
    sag-span ratio   52–4
    tower design   146, 207, 215
systematic design method   25, 27, 28

table of options   25, 28–33, 197
    car engine location   13–17, 28–30, 32–3
    wave energy converter   35–6
tankers, LNG   6, 23–5, 31, 33–4, 51–2
tankers, oil   50–1, 63
task-cost basis   16, 62, 179, 185
task measure   102
tasks as quantified functions   179
thermodynamic efficiency   89–91
tin-openers   45, 48
tolerances   60–2, 165
torsion rod   163–4
toy spade example   140
traffic in towns   97, 104, 105

transformers   11
turbines, electrical power   111–12
turbines, gas   3, 57–9, 198
    centring devices   133–4, 167–9
    cooling-air shroud   80–1
    parametric mapping   109–10
    scale effects   95–6
    see also   blade attachment, rotor; rotors
turbines, steam   82–3, 104, 106, 107–10, 112–13
turbo-alternators   111–12, 114
turning value problems   62–4
typewriters   44

value engineering   178
value (task ability)   179
values in gearing   180–4

Wankel engine   25
Watt's engine   10, 40, 82–3, 207
wave energy converters   34–9, 194–5, 199
    bobbing buoy   34–6
    Flounder   39
    Lancaster bag   37–8
    oscillating water column   36–7
    Salter's duck   37, 39
weight, minimum   3, 20, 42
weighting of options   30, 43
windlass   107